$$A = B$$

A = B

Marko Petkovšek
University of Ljubljana
Ljubljana, Slovenia

Herbert S. Wilf
University of Pennsylvania
Philadelphia, PA, USA

Doron Zeilberger
Temple University
Philadelphia, PA, USA

CRC Press
Taylor & Francis Group
Boca Raton London New York

CRC Press is an imprint of the
Taylor & Francis Group, an **informa** business

CRC Press
Taylor & Francis Group
6000 Broken Sound Parkway NW, Suite 300
Boca Raton, FL 33487-2742

© 1996 by Taylor & Francis Group, LLC
CRC Press is an imprint of Taylor & Francis Group, an Informa business

First issued in paperback 2019

No claim to original U.S. Government works

ISBN 13: 978-0-367-44871-4 (pbk)
ISBN 13: 978-1-56881-063-8 (hbk)

Visit the Taylor & Francis Web site at
http://www.taylorandfrancis.com

and the CRC Press Web site at
http://www.crcpress.com

[50] Develop computer programs for simplifying sums that involve binomial coefficients.

Exercise 1.2.6.63 in
The Art of Computer Programming, Volume 1: Fundamental Algorithms
by Donald E. Knuth,
Addison Wesley, Reading, Massachusetts, 1968.

Contents

Foreword

Science is what we understand well enough to explain to a computer. Art is everything else we do. During the past several years an important part of mathematics has been transformed from an Art to a Science: No longer do we need to get a brilliant insight in order to evaluate sums of binomial coefficients, and many similar formulas that arise frequently in practice; we can now follow a mechanical procedure and discover the answers quite systematically.

I fell in love with these procedures as soon as I learned them, because they worked for me immediately. Not only did they dispose of sums that I had wrestled with long and hard in the past, they also knocked off two new problems that I was working on at the time I first tried them. The success rate was astonishing.

In fact, like a child with a new toy, I can't resist mentioning how I used the new methods just yesterday. Long ago I had run into the sum $\sum_k \binom{2n-2k}{n-k}\binom{2k}{k}$, which takes the values 1, 4, 16, 64 for $n = 0, 1, 2, 3$ so it must be 4^n. Eventually I learned a tricky way to prove that it is, indeed, 4^n; but if I had known the methods in this book I could have proved the identity immediately. Yesterday I was working on a harder problem whose answer was $S_n = \sum_k \binom{2n-2k}{n-k}^2 \binom{2k}{k}^2$. I didn't recognize any pattern in the first values 1, 8, 88, 1088, so I computed away with the Gosper-Zeilberger algorithm. In a few minutes I learned that $n^3 S_n = 16(n - \frac{1}{2})(2n^2 - 2n + 1)S_{n-1} - 256(n - 1)^3 S_{n-2}$.

Notice that the algorithm doesn't just verify a conjectured identity "$A = B$". It also answers the question "What is A?", when we haven't been able to formulate a decent conjecture. The answer in the example just considered is a nonobvious recurrence from which it is possible to rule out any simple form for S_n.

I'm especially pleased to see the appearance of this book, because its authors have not only played key roles in the new developments, they are also master expositors of mathematics. It is always a treat to read their publications, especially when they are discussing really important stuff.

Science advances whenever an Art becomes a Science. And the state of the Art advances too, because people always leap into new territory once they have understood more about the old. This book will help you reach new frontiers.

Donald E. Knuth
Stanford University
20 May 1995

A Quick Start ...

You've been up all night working on your new theory, you found the answer, and it's in the form of a sum that involves factorials, binomial coefficients, and so on, such as

$$f(n) = \sum_{k=0}^{n} (-1)^k \binom{x-k+1}{k} \binom{x-2k}{n-k}.$$

You know that many sums like this one have simple evaluations and you would like to know, quite definitively, if this one does, or does not. Here's what to do.

1. Let $F(n,k)$ be your *summand*, i.e., the function[1] that is being summed. Your first task is to find the recurrence that F satisfies.

2. If you are using Mathematica, go to step 4 below. If you are using Maple, then get the package EKHAD either from the included diskette or from the WorldWideWeb site given on page 201. Read in EKHAD, and type

   $$\texttt{zeil(F(n,k),k,n,N);}$$

   in which your summand is typed, as an expression, in place of "F(n,k)". So in the example above you might type

   ```
   f:=(n,k)->(-1)^k*binomial(x-k+1,k)*binomial(x-2*k,n-k);
   zeil(f(n,k),k,n,N);
   ```

   Then zeil will print out the recurrence that your summand satisfies (it *does* satisfy one; see theorems 4.4.1 on page 65 and 6.2.1 on page 107). The output recurrence will look like eq. (6.1.3) on page 104. In this example zeil prints out the recurrence

   $$((n+2)(n-x) - (n+2)(n-x)N^2)F(n,k) = G(n,k+1) - G(n,k),$$

---
[1] But what is the little icon in the right margin? See page 9.

where N is the forward shift operator and G is a certain function that we will ignore for the moment. In customary mathematical notation, `zeil` will have found that

$$(n+2)(n-x)F(n,k) - (n+2)(n-x)F(n+2,k) = G(n,k+1) - G(n,k).$$

3. The next step is to sum the recurrence that you just found over all the values of k that interest you. In this case you can sum over all integers k. The right side telescopes to zero, and you end up with the recurrence that your *unknown sum* $f(n)$ satisfies, in the form

$$f(n) - f(n+2) = 0.$$

Since $f(0) = 1$ and $f(1) = 0$, you have found that $f(n) = 1$, if n is even, and $f(n) = 0$, if n is odd, and you're all finished. If, on the other hand, you get a recurrence whose solution is not obvious to you because it is of order higher than the first and it does not have constant coefficients, for instance, then go to step 5 below.

4. If you are using Mathematica, then get the program Zb (see page 116 below) in the package **paule-schorn** from the WorldWideWeb site given on page 201. Read in Zb, and type

```
Zb[(-1)^k Binomial[x-k+1,k] Binomial[x-2k,n-k],k,n,1]
```

in which the final "1" means that you are looking for a recurrence of order 1. In this case the program will not find a recurrence of order 1, and will type "**try higher order.**" So rerun the program with the final "1" changed to a "2". Now it will find the same recurrence as in step 2 above, so continue as in step 3 above.

5. If instead of the easy recurrence above, you got one of higher order, and with polynomial-in-n coefficients, then you will need algorithm Hyper, on page 156 below, to solve it for you, or to *prove* that it cannot be solved in closed form (see page 145 for a definition of "closed form"). This program is also on the diskette that came with this book, or it can be downloaded from the WWW site given on page 201. Use it just as in the examples in Section 8.5. You are guaranteed either to find the closed form evaluation that you wanted, or else to find a proof that none exists.

Part I

Background

Chapter 1

Proof Machines

The ultimate goal of mathematics is to eliminate any need for intelligent thought.

—Alfred N. Whitehead

1.1   Evolution of the province of human thought

One of the major themes of the past century has been the growing replacement of human thought by computer programs. Whole areas of business, scientific, medical, and governmental activities are now computerized, including sectors that we humans had thought belonged exclusively to us. The interpretation of electrocardiogram readings, for instance, can be carried out with very high reliability by software, without the intervention of physicians—not perfectly, to be sure, but very well indeed. Computers can fly airplanes; they can supervise and execute manufacturing processes, diagnose illnesses, play music, publish journals, etc.

The frontiers of human thought are being pushed back by automated processes, forcing people, in many cases, to relinquish what they had previously been doing, and what they had previously regarded as their safe territory, but hopefully at the same time encouraging them to find new spheres of contemplation that are in no way threatened by computers.

We have one more such story to tell in this book. It is about discovering new ways of finding beautiful mathematical relations called identities, and about proving ones that we already know.

People have always perceived and savored *relations* between natural phenomena. First these relations were *qualitative*, but many of them sooner or later became *quantitative*. Most (but not all) of these relations turned out to be *identities*, that is, statements whose format is $\mathbf{A} = \mathbf{B}$, where $\mathbf{A}$ is one quantity and $\mathbf{B}$ is another quantity, and the surprising fact is that they are really the same.

Before going on, let's recall some of the more celebrated ones:

- $a^2 + b^2 = c^2$.

- When Archimedes (or, for that matter, you or I) takes a bath, it happens that "Loss of Weight" = "Weight of Fluid Displaced."

- $a(\frac{-b \pm \sqrt{b^2 - 4ac}}{2a})^2 + b(\frac{-b \pm \sqrt{b^2 - 4ac}}{2a}) + c = 0$.

- $F = ma$.

- $V - E + F = 2$.

- $\det(AB) = \det(A)\det(B)$.

- $\mathbf{curl\ H} = \frac{\partial \mathbf{D}}{\partial t} + j \qquad \mathbf{div \cdot B} = 0 \qquad \mathbf{curl\ E} = -\frac{\partial \mathbf{B}}{\partial t} \qquad \mathbf{div \cdot D} = \rho$.

- $E = mc^2$.

- Analytic Index = Topological Index.   (The Atiyah–Singer theorem)

- The cardinality of $\{x, y, z, n \in \mathbb{Z} | xyz \neq 0, n > 2, x^n + y^n = z^n\} = 0$.

As civilization grew older and (hopefully) wiser, it became not enough to *know* the facts, but instead it became necessary to *understand* them as well, and to *know for sure*. Thus was born, more than 2300 years ago, the notion of *proof*. Euclid and his contemporaries tried, and partially succeeded in, *deducing* all facts about plane geometry from a certain number of *self-evident* facts that they called *axioms*. As we all know, there was one axiom that turned out to be not as self-evident as the others: the notorious *parallel* axiom. Liters of ink, kilometers of parchment, and countless feathers were wasted trying to show that it is a theorem rather than an axiom, until Bolyai and Lobachevski shattered this hope and showed that the parallel axiom, in spite of its lack of self-evidency, is a genuine axiom.

Self-evident or not, it was still tacitly assumed that all of mathematics was recursively axiomatizable, i.e., that every conceivable truth could be deduced from some set of axioms. It was David Hilbert who, about 2200 years after Euclid's death, wanted a *proof* that this is indeed the case. As we all know, but many of us choose to ignore, this tacit assumption, made explicit by Hilbert, turned out to be false. In 1930, 24-year-old Kurt Gödel proved, using some ideas that were older than Euclid, that no matter how many axioms you have, as long as they are not contradictory there will always be some facts that are *not* deducible from the axioms, thus delivering another blow to overly simple views of the complex texture of mathematics.

Closely related to the activity of *proving* is that of *solving*. Even the ancients knew that not all equations have solutions; for example, the equations $x + 2 = 1$, $x^2 + 1 = 0$, $x^5 + 2x + 1 = 0$, $P = \neg P$, have been, at various times, regarded as being of that kind. It would still be nice to *know*, however, whether our failure to find a solution is intrinsic or due to our incompetence. Another problem of Hilbert was to *devise a process according to which it can be determined by a finite number of operations whether a [diophantine] equation is solvable in rational integers*. This dream was also shattered. Relying on the seminal work of Julia Robinson, Martin Davis, and Hilary Putnam, 22-year-old Yuri Matiyasevich proved [Mati70], in 1970, that such a "process" (which nowadays we call an *algorithm*) does not exist.

What about identities? Although theorems and diophantine equations are undecidable, mightn't there be at least a *Universal Proof Machine* for humble statements like $A = B$? Sorry folks, no such luck.

Consider the identity

$$\sin^2(|(\ln 2 + \pi x)^2|) + \cos^2(|(\ln 2 + \pi x)^2|) = 1.$$

We leave it as an exercise for the reader to prove. However, not all such identities are decidable. More precisely, we have Richardson's theorem ([Rich68], see also [Cavi70]).

Theorem 1.1.1 *(Richardson) Let $\mathcal{R}$ consist of the class of expressions generated by*

1. *the rational numbers and the two real numbers π and $\ln 2$,*

2. *the variable x,*

3. *the operations of addition, multiplication, and composition, and*

4. *the sine, exponential, and absolute value functions.*

If $E \in \mathcal{R}$, the predicate "$E = 0$" is recursively undecidable.

A pessimist (or, depending on your point of view, an optimist) might take all these negative results to mean that we should abandon the search for "Proof Machines" altogether, and be content with proving one identity (or theorem) at a time. Our \$5 pocket calculator shows that this is nonsense. Suppose we have to prove that $3 \times 3 = 9$. A rigorous but ad hoc proof goes as follows. By definition $3 = 1 + 1 + 1$. Also by definition, $3 \times 3 = 3 + 3 + 3$. Hence $3 \times 3 = (1 +$

$1 + 1) + (1 + 1 + 1) + (1 + 1 + 1)$, which by the associativity of addition, equals $1 + 1 + 1 + 1 + 1 + 1 + 1 + 1 + 1$, which *by definition* equals 9.   □

However, thanks to the Indians, the Arabs, Fibonacci, and others, there is a *decision procedure* for deciding all such numerical identities involving integers and using addition, subtraction, and multiplication. Even more is true. There is a *canonical form* (the decimal, binary, or even unary representation) to which every such expression can be reduced, and hence it makes sense to talk about *evaluating* such expressions in *closed form* (see page 145). So, not only can we decide whether or not $4 \times 5 = 20$ is true or false, we can *evaluate* the left hand side, and find out that it is 20, even without knowing the conjectured answer beforehand.

Let's give the floor to Dave Bressoud [Bres93]:

> "The existence of the computer is giving impetus to the discovery of algorithms that generate proofs. I can still hear the echoes of the collective sigh of relief that greeted the announcement in 1970 that there is no general algorithm to test for integer solutions to polynomial Diophantine equations; Hilbert's tenth problem has no solution. Yet, as I look at my own field, I see that creating algorithms that generate proofs constitutes some of the most important mathematics being done. The all-purpose proof machine may be dead, but tightly targeted machines are thriving."

In this book we will describe in detail several such *tightly targeted machines*. Our main targets will be *binomial coefficient identities, multiple hypergeometric (and more generally, holonomic) integral/sum identities*, and *q-identities*. In dealing with these subjects we will for the most part discuss in detail only single-variable non-*q* identities, while citing the literature for the analogous results in more general situations. We believe that these are just modest first steps, and that in the future we, or at least our children, will witness many other such targeted proof machines, for much more general classes, or completely different classes, of identities and theorems. Some of the more plausible candidates for the near future are described in Chapter 9 . In the rest of this chapter, we will briefly outline some older proof machines. Some of them, like that for adding and multiplying integers, are very well known. Others, such as the one for *trigonometric identities*, are well known, but not as well known as they should be. Our poor students are still asked to prove, for example, that $\cos 2x = \cos^2 x - \sin^2 x$. Others, like identities for elliptic functions, were perhaps only implicitly known to be routinely provable, and their routineness will be pointed out explicitly for the first time here.

The key for designing proof machines for classes of identities is that of finding a *canonical form*, or failing this, finding at least a *normal form*.

1.2 Canonical and normal forms

Canonical forms

Given a set of objects (for example, people), there may be many ways to describe a particular object. For example "Bill Clinton" and "the president of the USA in 1995," are two descriptions of the same object. The second one defines it uniquely, while the first one most likely doesn't. Neither of them is a good canonical form. A *canonical form* is a clear-cut way of describing *every* object in the class, in a one-to-one way. So in order to find out whether object A equals object B, all we have to do is find their canonical forms, $c(A)$ and $c(B)$, and check whether or not $c(A)$ equals $c(B)$.

Example 1.2.1. Prove the following identity

The Third Author of This Book = The Prover of the Alternating Sign Matrix Conjecture [Zeil96a].

Solution: First verify that both sides of the identity are objects that belong to a well-defined class that possesses a canonical form. In this case the class is that of citizens of the USA, and a good canonical form is the Social Security number. Next compute (or look up) the Social Security Number of both sides of the equation. The SSN of the left side is 555123456. Similarly, the SSN of the right side is[1] 555123456. Since the canonical forms match, we have that, indeed, $A = B$. □

Another example is $5 + 7 = 3 + 9$. Both sides are integers. Using the decimal representation, the canonical forms of both sides turn out to be $1 \cdot 10^1 + 2 \cdot 10^0$. Hence the two sides are equal.

Normal forms

So far, we have not assumed anything about our set of objects. In the vast majority of cases in mathematics, the set of objects will have at least the structure of an *additive group*, which means that you can add and, more importantly, subtract. In such cases, in order to prove that $A = B$, we can prove the equivalent statement $A - B = 0$. A *normal form* is a way of representing objects such that although an object may have many "names" (i.e., $c(A)$ is a set), every possible name corresponds to exactly one object. In particular, you can tell right away whether it represents 0. For example, every rational number can be written as a quotient of integers a/b, but in many ways. So $15/10$ and $30/20$ represent the same entity. Recall that the

---

[1] Number altered to protect the innocent.

set of rational numbers is equipped with addition and subtraction, given by

$$\frac{a}{b} + \frac{c}{d} = \frac{ad + bc}{bd}, \quad \frac{a}{b} - \frac{c}{d} = \frac{ad - bc}{bd}.$$

How can we prove an identity such as $13/10 + 1/5 = 29/20 + 1/20$? All we have to do is prove the equivalent identity $13/10 + 1/5 - (29/20 + 1/20) = 0$. The left side equals $0/20$. We know that any fraction whose numerator is 0 stands for 0. The proof machine for proving numerical identities $A = B$ involving rational numbers is thus to compute *some* normal form for $A - B$, and then check whether the numerator equals 0.

The reader who prefers canonical forms might remark that rational numbers *do* have a canonical form: a/b with a and b relatively prime. So another algorithm for proving $A = B$ is to compute normal forms for both A and B, then, by using the Euclidean algorithm, to find the GCD of numerator and denominator on both sides, and cancel out by them, thereby reducing both sides to "canonical form."

1.3   Polynomial identities

Back in ninth grade, we were fascinated by formulas like $(x+y)^2 = x^2 + 2xy + y^2$. It seemed to us to be of such astounding generality. No matter what numerical values we would plug in for x and y, we would find that the left side equals the right side. Of course, to our jaded contemporary eyes, this seems to be as routine as $2+2 = 4$. Let us try to explain *why*. The reason is that both sides are polynomials in the two variables x, y. Such polynomials have a canonical form

$$P = \sum_{i \geq 0, j \geq 0} a_{i,j} x^i y^j,$$

where only finitely many $a_{i,j}$ are non-zero.

The Maple function expand translates polynomials to normal form (though one might insist that $x^2 + y$ and $y + x^2$ look different, hence this is really a normal form only). Indeed, the easiest way to prove that $A = B$ is to do expand(A-B) and see whether or not Maple gives the answer 0.

Even though they are completely routine, polynomial identities (and by clearing denominators, also identities between rational functions) can be very important. Here are some celebrated ones:

$$\left(\frac{a+b}{2}\right)^2 - ab = \left(\frac{a-b}{2}\right)^2, \tag{1.3.1}$$

which immediately implies the arithmetic-geometric-mean inequality; Euler's

$$(a^2 + b^2 + c^2 + d^2)(A^2 + B^2 + C^2 + D^2) =$$
$$(aA + bB + cC + dD)^2 + (aB - bA - cD + dC)^2$$
$$+ (aC + bD - cA - dB)^2 + (aD - bC + cB - dA)^2, \quad (1.3.2)$$

which shows that in order to prove that every integer is a sum of four squares it suffices to prove it for primes; and

$$(a_1^2 + a_2^2)(b_1^2 + b_2^2) - (a_1 b_1 + a_2 b_2)^2 = (a_1 b_2 - a_2 b_1)^2,$$

which immediately implies the Cauchy-Schwarz inequality in two dimensions.

About our terminal logos:

Throughout this book, whenever you see the computer terminal logo in the margin, like this, and if its screen is white, it means that we are about to do something that is very computer-ish, so the material that follows can be either skipped, if you're mainly interested in the mathematics, or especially savored, if you are a computer type.

When the computer terminal logo appears with a darkened screen, the normal mathematical flow will resume, at which point you may either resume reading, or flee to the next terminal logo, again depending, respectively, on your proclivities.

1.4   Proofs by example?

Are the following proofs acceptable?

Theorem 1.4.1 *For all integers $n \geq 0$,*

$$\sum_{i=1}^{n} i^3 = \left(\frac{n(n+1)}{2} \right)^2.$$

Proof. For $n = 0, 1, 2, 3, 4$ we compute the left side and fit a polynomial of degree 4 to it, viz. the right side. □

Theorem 1.4.2 *For every triangle ABC, the angle bisectors intersect at one point.*

Proof. Verify this for the 64 triangles for which $\angle A = 10°, 20°, \ldots, 80°$ and $\angle B = 10°, 20°, \ldots, 80°$. Since the theorem is true in these cases it is always true. □

If a student were to present these "proofs" you would probably fail him. We won't. The above proofs are completely rigorous. To make them more readable, one may add, in the first proof, the phrase: "Both sides obviously satisfy the relations $p(n) - p(n-1) = n^3$; $p(0) = 0$," and in the second proof: "It is easy to see that the coordinates of the intersections of the pairs of angle bisectors are rational functions of degrees ≤ 7 in $a = \tan(\angle A/2)$ and $b = \tan(\angle B/2)$. Hence if they agree at 64 points (a, b), they are identical."

The principle behind these proofs is that if our set of objects consists of polynomials $p(n)$ of degree $\leq L$ in n, for a fixed L, then for every distinct set of inputs, say $\{0, 1, \ldots, L\}$, the vector $c(p) = [p(0), p(1), \ldots, p(L)]$ constitutes a *canonical form*. In practice, however, to prove a polynomial identity it is just as easy to expand the polynomials as explained above. Note that every identity of the form $\sum_{i=1}^{n} q(i) = p(n)$ is equivalent to the two routinely verifiable statements

$$p(n) - p(n-1) = q(n) \quad \text{and} \quad p(0) = 0.$$

A complete computer-era proof of Theorem 1.4.1 would go like this: Begin by suspecting that the sum of the first n cubes might be a fourth degree polynomial in n. Then use your computer to fit a fourth degree polynomial to the data points $(0, 0)$, $(1, 1)$, $(2, 9)$, $(3, 36)$, and $(4, 100)$. This polynomial will turn out to be

$$p(n) = (n(n+1)/2)^2.$$

Now use your computer algebra program to check that $p(n) - p(n-1) - n^3$ is the zero polynomial, and that $p(0) = 0$. $\qquad\qquad\square$

Theorem 1.4.2 is an example of a theorem in plane geometry. The fact that all such theorems are routine, at least in principle, has been known since René Descartes. Thanks to modern computer algebra systems, they are also routine in practice. More sophisticated theorems may need Buchberger's method of Gröbner bases [Buch76], which is also implemented in Maple, but for which there exists a *targeted* implementation by the computer algebra system *Macaulay* [BaySti] (see also [Davi95], and [Chou88]).

Here is the Maple code for proving Theorem 1.4.2 above.

```
#begin Maple Code
f:=proc(ta,tb):(ta+tb)/(1-ta*tb):end:
f2:=proc(ta);normal(f(ta,ta)):end:
anglebis:=proc(ta,tb):
eq1:=y=x*ta: eq2:=y=(x-1)*(-tb):
Eq1:=y=x*f2(ta):
Eq2:=y=(x-1)*(-f2(tb)):
sol:=solve({Eq1,Eq2},{x,y}):
Cx:=subs(sol,x):Cy:=subs(sol,y):
sol:=solve({eq1,eq2},{x,y}):
```

```
ABx:=subs(sol,x):ABy:=subs(sol,y):
eq3:=(y-Cy)=(x-Cx)*(-1/f(ta,-tb)):
sol:=solve({eq1,eq3},{x,y}):
ACx:=subs(sol,x):ACy:=subs(sol,y):
print(normal(ACx),normal(ABx)):
print(normal(ACy),normal(ABy)):
normal(ACx -ABx),normal(ACy-ABy);
end:
#end Maple code
```

To prove Theorem 1.4.2, all you have to do, after typing the above in a Maple session, is type `anglebis(ta,tb);`, and if you get $0,0$, you will have proved the theorem.

Let's briefly explain the program. W.l.o.g. $A = (0,0)$, and $B = (1,0)$. Call $\angle A = 2a$, and $\angle B = 2b$. The inputs are $ta := \tan a$ and $tb := \tan b$. All quantities are expressed in terms of ta and tb and are easily seen to be rational functions in them. The procedure `f(ta,tb)` implements the addition law for the tangent function:

$$\tan(a + b) = (\tan a + \tan b)/(1 - \tan a \tan b);$$

the variables `eq1`, `eq2`, `eq3` are the equations of the angle bisectors at A, B, and C respectively. (ABx, ABy) and (ACx, ACy) are the points of intersection of the bisectors of $\angle A$ and $\angle B$, and of $\angle A$ and $\angle C$, respectively, and the output, the last line, gives the differences. It should be $0,0$.

In *Plane Geometry: An Elementary Textbook* by Shalosh B. Ekhad, XIV, at `http://www.math.rutgers.edu/~zeilberg/GT.html` there are Maple proofs of Pascal's hexagon theorem, Morley's trisectors theorem, and many more theorems of plane geometry.

1.5 Trigonometric identities

The verification of any finite identity between trigonometric functions that involves only the four basic operations (not compositions!), where the arguments are of the form ax, for specific a's, is purely routine.

- **First Way:** Let $w := \exp(ix)$, then $\cos x = (w + w^{-1})/2$ and $\sin x = (w - w^{-1})/(2i)$. So equality of rational expressions in trigonometric functions can be reduced to equality of polynomial expressions in w. (**Exercise:** Prove, in this way, that $\sin 2x = 2 \sin x \cos x$.)

- **Second Way:** Whenever you see $\cos w$, change it to $\sqrt{1 - \sin^2 w}$, then replace $\sin w$, by z, say, then express everything in terms of arcsin. To prove the resulting identity, differentiate it with respect to one of the variables, and use the defining properties $\arcsin(z)' = (1 - z^2)^{-1/2}$, and $\arcsin(0) = 0$.

Example 1.5.1. By setting $\sin a = x$ and $\sin b = y$, we see that the identity $\sin(a + b) = \sin a \cos b + \sin b \cos a$ is equivalent to

$$\arcsin x + \arcsin y = \arcsin(x\sqrt{1 - y^2} + y\sqrt{1 - x^2}).$$

When $x = 0$ this is tautologous, so it suffices to prove that the derivatives of both sides with respect to x are the same. This is a routinely verifiable algebraic identity.

Below is the short Maple Code that proves it. If its output is zero then the identity has been proved.

```
f:=arcsin(x) + arcsin(y) :
g:= arcsin(x*(1-y**2)**( 1/2) + y*(1-x**2)**(1/2));
f1:=diff(f,x):   g1:=diff(g,x):
normal(simplify(expand(g1**2))-f1**2);
```

□

1.6   Fibonacci identities

All Fibonacci number identities such as Cassini's $F_{n+1}F_{n-1} - F_n^2 = (-1)^n$ (and much more complicated ones), are routinely provable using Binet's formula:

$$F_n := \frac{1}{\sqrt{5}} \left(\left(\frac{1 + \sqrt{5}}{2} \right)^n - \left(\frac{1 - \sqrt{5}}{2} \right)^n \right).$$

Below is the Maple code that proves Cassini's formula.

```
F:=proc(n):
(((1+sqrt(5))/2)**n-((1-sqrt(5))/2)**n)/sqrt(5):
end:
Cas:=F(n+1)*F(n-1)-F(n)**2:
Cas:=expand(simplify(Cas)):
numer(Cas)/expand(denom(Cas));
```

1.7   Symmetric function identities

Consider the identity

$$\left(\sum_{i=1}^{n} a_i \right)^2 = \sum_{i=1}^{n} a_i^2 + 2 \sum_{1 \leq i < j \leq n} a_i a_j,$$

where n is an *arbitrary* integer. Of course, for every *fixed* n, no matter how big, the above is a routine polynomial identity. We claim that it is purely routine, even for arbitrary n, and that in order to verify it we can take, without loss of generality, $n = 2$. The reason is that both sides are *symmetric functions*, and denoting, as usual,

$$p_k := \sum_{i=1}^{n} a_i^k, \quad e_k := \sum_{1 \le i_1 < \cdots < i_k \le n} a_{i_1} \cdots a_{i_k},$$

the above identity can be rephrased as

$$p_1^2 = p_2 + 2e_2.$$

Now it follows from the theory of symmetric functions (e.g., [Macd95]) that every polynomial identity between the e_i's and p_i's (and the other bases for the space of symmetric functions as well) is purely routine, and is true if and only if it is true for a certain finite value of n, namely the largest index that shows up in the e's and p's. This is also true if we have several sets of variables, $a_i, b_i \ldots$, and by 'symmetric' we mean that the polynomial remains unchanged when we simultaneously permute the a_i's, b_i's, and so on. Thus the following identity, which implies the Cauchy-Schwarz inequality for *every* dimension, is also routine:

$$\sum_{i=1}^{n} a_i^2 \sum_{i=1}^{n} b_i^2 - (\sum_{i=1}^{n} a_i b_i)^2 = \sum_{1 \le i < j \le n} (a_i b_j - a_j b_i)^2. \qquad (1.7.1)$$

For the study of symmetric functions we highly recommend John Stembridge's Maple package SF, which is available by ftp to ftp.math.lsa.umich.edu.

1.8 Elliptic function identities

One must [not] always invert

— Carl G. J. Jacobi [Shalosh B. Ekhad]

It is lucky that computers had not yet been invented in Jacobi's time. It is possible that they would have prevented the discovery of one of the most beautiful theories in the whole of mathematics: the theory of elliptic functions, which leads naturally to the theory of modular forms, and which, besides being gorgeous for its own sake [Knop93], has been applied all over mathematics (e.g., [Sarn93]), and was crucial in Wiles's proof of Fermat's last theorem.

Let's engage in a bit of revisionist history. Suppose that the trigonometric functions had not been known before calculus. Then in order to find the perimeter of a quarter-circle, we would have had to evaluate:

$$\int_0^1 \sqrt{1 + \left(\frac{dy}{dx}\right)^2},$$

where $y = \sqrt{1 - x^2}$. This turns out to be

$$\int_0^1 \frac{dx}{\sqrt{1 - x^2}}, \qquad (1.8.1)$$

which may be taken as the *definition* of $\pi/2$. We can call this the *complete circular integral*. More generally, suppose that we want to know $F(z)$, the arc length of the circle above the interval $[0, z]$, for general z. Then the integral is the *incomplete circular integral*

$$F(z) := \int_0^z \frac{dx}{\sqrt{1 - x^2}}, \qquad (1.8.2)$$

which may also be defined by $F'(z) = (1 - z^2)^{-1/2}$, $F(0) = 0$. Then it is possible that some genius would have come up with the idea of defining $\sin w := F^{-1}(w)$, $\cos w := F^{-1}(\pi/2 - w)$, and realized that $\sin z$ and $\cos z$ are much easier to handle, and to compute with, than $\arcsin z$. Furthermore, that genius would have soon realized how to express the sine and cosine functions in terms of the exponential function. Using its Taylor expansion, which converges very rapidly, the aforementioned genius would have been able to compile a table of the sine function, from which automatically would have resulted a table of the function of primary interest, $F(z)$ above (which in real life is called "arcsine" or the "inverse sine" function.)

Now let's go back to real history. Consider the analogous problem for the arc length of the ellipse. This involves an integral of the form

$$F(z) := \int_0^z \frac{dx}{\sqrt{(1 - x^2)(1 - k^2 x^2)}}, \qquad (1.8.3)$$

where k is a parameter $\in [0, 1]$. The study of these integrals was at the frontier of mathematical research in the first half of the nineteenth century. Legendre struggled with them for a long time, and must have been frustrated when Jacobi had the great idea of *inverting* $F(z)$. In analogy with the sine function, Jacobi called $F^{-1}(w)$, $\mathrm{sn}(w)$, and also defined $\mathrm{cn}(w) := \sqrt{1 - \mathrm{sn}^2(w)}$, and $\mathrm{dn}(w) := \sqrt{1 - k^2 \mathrm{sn}^2(w)}$. These are the (once) famous Jacobi elliptic functions. Jacobi realized that the counterparts of the exponential function are the so-called Jacobi theta functions, and he was able to express his elliptic functions in terms of his theta functions. His theta functions, one of which is

$$\theta_3(z) = 1 + 2\sum_{n=1}^{\infty} q^{n^2} \cos(2nz),$$

have series which converge very rapidly when q is small. With the aid of his famous transformation formula (see, e.g., [Bell61]) he was always able to compute his theta functions with very rapidly converging series. This enabled him (or his human computers) to compile highly accurate tables of his elliptic functions, and hence, of course, of the incomplete elliptic integral $F(z)$. Much more importantly, it led to a beautiful theory, which is still flourishing.

If Legendre's and Jacobi's contemporaries had had computers, it would have been relatively easy for them to have used numerical integration in order to compile a table of $F(z)$, and most of the motivation to *invert* would have gone. Had they had computer algebra, they would have also realized that all identities between elliptic functions are routine, and that it is not necessary to introduce theta functions. Take for example the addition formula for $\mathrm{sn}(w)$ (e.g., [Rain60], p. 348):

$$\mathrm{sn}(u+v) = \frac{\mathrm{sn}(u)\,\mathrm{cn}(v)\,\mathrm{dn}(v) + \mathrm{sn}(v)\,\mathrm{cn}(u)\,\mathrm{dn}(u)}{1 - k^2\,\mathrm{sn}^2(u)\,\mathrm{sn}^2(v)}. \qquad (1.8.4)$$

Putting $\mathrm{sn}(u) = x$, $\mathrm{sn}(v) = y$, and denoting, as above, sn^{-1} by F, we have that (1.8.4) is equivalent to

$$F(x) + F(y) = F\left(\frac{x\sqrt{1-y^2}\sqrt{1-k^2y^2} + y\sqrt{1-x^2}\sqrt{1-k^2x^2}}{1 - k^2x^2y^2}\right). \qquad (1.8.5)$$

This is routine. Indeed, when $x = 0$, both sides equal $F(y)$, and differentiating both sides with respect to x, using the chain rule and the defining property

$$F'(z) = \frac{1}{\sqrt{(1-z^2)(1-k^2z^2)}},$$

we get a finite algebraic identity.

If the following Maple code outputs a 1 (it did for us) then it would be a completely rigorous proof of Jacobi's addition formula for the *sn* function. Try to work this out by hand, and see that it would have been a formidable task for any human, even a Jacobi or Legendre.

```
lef:=F(x)+F(y):
rig:=
```

```
F((x *sqrt(1-y**2)*sqrt(1-k**2*y**2)
+y*sqrt(1-x**2)*sqrt(1-k**2*x**2))/(1-k**2*x**2*y**2)):
lef1:=1/sqrt((1-x**2)*(1-k**2*x**2)):rig1:=diff(rig,x);
g:=z->1/sqrt(1-z**2)/sqrt(1-k**2*z**2):
rig1:=subs(D(F)=g,rig1);
gu:=normal((rig1/lef1)**2);
expand(numer(gu))/expand(denom(gu));
```

Chapter 2

Tightening the Target

2.1   Introduction

In the next several chapters we are going to narrow the focus of the discussion from the whole world of identities to the kind of identities that tend to occur in combinatorial mathematics: *hypergeometric identities*. These are relations in which typically a sum of some huge expression involving binomial coefficients, factorials, rational functions and power functions is evaluated, and it miraculously turns out to be something very simple.

We will show you how to evaluate and to prove such sums entirely mechanically, i.e., "no thought required." Your computer will do the work. Everybody knows that computers are fast. In this book we'll try to show you that in at least one field of mathematics they are not only fast but smart, too.

What that means is that they can find very pretty proofs of very difficult theorems in the field of combinatorial identities. The computers do that by themselves, unassisted by hints or nudges from humans.

It means also that not only can your PC find such a proof, but you will be able to check the proof easily. So you won't have to take the computer's word for it. That is a very important point. People get unhappy when a computer blinks its lights for a while and then announces a result, if people cannot easily check the truth of the result for themselves. In this book you will be pleased to note that although the computers will have to blink their lights for quite a long time, when they are finished they will give to us people a short certificate from which it will be easy to check the truth of what they are claiming.

Computers not only find proofs of known identities, they also find completely new identities. Lots of them. Some very pretty. Some not so pretty but very useful. Some neither pretty nor useful, in which case we humans can ignore them.

The body of work that has resulted in these automatic "summation machines" is very recent, and it has had contributions from several researchers. Our discussion will be principally based on the following:

- [Fase45] is the Ph.D. dissertation of Sister Mary Celine Fasenmyer, in 1945. It showed how recurrences for certain polynomial sequences could be found algorithmically. (See Chapter 4.)

- [Gosp78], by R. W. Gosper, Jr., is the discovery of the algorithmic solution of the problem of indefinite hypergeometric summation (see Chapter 5). Such a summation is of the form $f(n) = \sum_{k=0}^{n} F(k)$, where F is hypergeometric.

- [Zeil82], of Zeilberger, recognized that Sister Celine's method would also be the basis for proving combinatorial identities by recurrence. (See Chapter 4.)

- [Zeil91, Zeil90b], also by Zeilberger, developed his "creative telescoping" algorithm for finding recurrences for combinatorial summands, which greatly accelerated the one of Sister Celine. (See Chapter 6.)

- [WZ90a], of Wilf and Zeilberger, finds a special case of the above which enables the discovery of new identities from old as well as very short and elegant proofs. (See Chapter 7.)

- [WZ92a], also by Wilf and Zeilberger, generalizes the methods to multisums, q-sums, etc., as well as giving proofs of the fundamental theorems and explicit estimates for the orders of the recurrences involved. (See Chapter 4.)

- [Petk91] is the Ph.D. thesis of Marko Petkovšek, in 1991. In it he discovered the algorithm for deciding if a given recurrence with polynomial coefficients has a "simple" solution, which, together with the algorithms above, enables the automated discovery of the simple evaluation of a given *definite* sum, if one exists, or a proof of nonexistence, if none exists (see Chapter 8). A *definite* hypergeometric sum is one of the form $f(n) = \sum_{k=-\infty}^{\infty} F(n, k)$, where F is hypergeometric.

Suppose you encounter a large sum of factorials and binomial coefficients and whatnot. You would like to know whether or not that sum can be expressed in a much simpler way, say as a single term that involves factorials, etc. In this book we will show you how several recently developed computer algorithms can do the job for you. If there is a simple form, the algorithms will find it. If there isn't, they will prove that there isn't.

In fact, the previous paragraph is probably the most important single message of this book, so we'll say it again:

> *The problem of discovering whether or not a given hypergeometric sum is express-ible in simple "closed form," and if so, finding that form, and if not, proving that it is not, is a task that computers can now carry out by themselves, with guar-anteed success under mild hypotheses about what a "hypergeometric term" is (see Section 4.4) and what a "closed form" is (See page 145, where it is essentially defined to mean a linear combination of a fixed number of hypergeometric terms).*

So if you have been working on some kind of mammoth sum or multiple sum, and have been searching for ways to simplify it, after long hours of fruitless labor you might feel a little better if you could be told that the sum simply *can't* be simplified. Then at least you would know that it wasn't your fault. Nobody will ever be able to simplify that expression, within a certain set of conventions about what simplification means, anyway.

We will present the underlying mathematical theory of these methods, the prin-cipal theorems and their proofs, and we also include a package of computer programs that will do these tasks (see Appendix A).

The main theme that runs through these methods is that of *recurrence*. To find out if a sum can be simplified, we find a recurrence that the sum satisfies, and we then either solve the recurrence explicitly, or else prove that it can't be solved explicitly, under a very reasonable definition of "explicit." Your computer will find the recurrence that a sum satisfies (see Chapter 6), and then decide if it can be solved in a simple form (see Chapter 8).

For instance, a famous old identity states that the sum of all of the binomial coefficients of a given order n is 2^n. That is, we have

$$\sum_k \binom{n}{k} = 2^n.$$

The sum of the squares of the binomial coefficients is something simple, too:

$$\sum_k \binom{n}{k}^2 = \binom{2n}{n}.$$

Range convention: Please note that, throughout this book, when ranges of summation are not specified, then the sums are understood to extend over *all integers*, positive and negative. In the above sum, for instance, the binomial coefficient $\binom{n}{k}$ vanishes if $k < 0$ or $k > n \geq 0$ (assuming n is an integer), so only finitely many terms contribute.

But what about the sum of their *cubes*? For many years people had searched for a simple formula in this case and hadn't found one. Now, thanks to newly developed computer methods, it can be proved that no "simple" formula exists. This is done by finding a recurrence formula that the sum of the cubes satisfies and then showing that the recurrence has no "simple" solution (see Theorem 8.8.1 on page 164).

The definition of the term "simple formula" will be made quite precise when we discuss this topic in more depth in Chapter 8. By the way, a recurrence that $f(n) = \sum_k \binom{n}{k}^3$ satisfies turns out to be

$$8(n+1)^2 f(n) + (7n^2 + 21n + 16) f(n+1) - (n+2)^2 f(n+2) = 0.$$

Your computer will find that for you. All you have to do is type in the summand $\binom{n}{k}^3$. After finding the recurrence your computer will then prove that it has no solution in "closed form," in a certain precise sense[1].

Would you like to know how all of that is done? Read on.

The sum $\sum_k \binom{n}{k}^3$, of course, is just one of many examples of formulas that can be treated with these methods.

If you aren't interested in finding or proving an identity, you might well be interested in finding a *recurrence* relation that an unknown sum satisfies. Or in deciding whether a given linear recurrence relation with polynomial coefficients can be solved in some explicit way. In that case this book has some powerful tools for you to use.

This book contains both mathematics and software, the former being the theoretical underpinnings of the latter. For those who have not previously used them, the programs will likely be a revelation. Imagine the convenience of being able to input a sum that one is interested in and having the program print out a simple formula that evaluates it! Think also of inputting a complicated sum and getting a recurrence formula that it satisfies, automatically.

We hope you'll enjoy both the mathematics and the software. Taken together, they are the story of a sequence of very recent developments that have changed the field on which the game of discrete mathematics is played.

We think about identities a little differently in this book. The computer methods tend in certain directions that seem not to come naturally to humans. We illustrate the thought processes by a small example.

Example 2.1.1. Define $e(x)$ to be the famous series $\sum_{n \geq 0} x^n / n!$. We will prove that $e(x + y) = e(x)e(y)$ for all x and y.

First, the series converges for all x, by the ratio test, so $e(x)$ is well defined for all x, and $e'(x) = e(x)$. Next, instead of trying to prove that the two sides of the

---

[1]See page 145.

identity are equal, let's prove that their ratio is 1 (that will be a frequent tactic in this book). Not only that, we'll prove that the ratio is 1 by differentiating it and getting 0 (another common tactic here).

So define the function $F(x, y) = e(x + y)e(-x)e(-y)$. By direct differentiation we find that $D_x F = D_y F = 0$. Thus F is constant. Set $x = y = 0$ to find that the constant is 1. Thus $e(x + y)e(-x)e(-y) = 1$ for all x, y. Now let $y = 0$ to find that $e(-x) = 1/e(x)$. Thus $e(x + y) = e(x)e(y)$ for all x, y, as claimed. $\qquad\square$

We urge you to have available one of several commercially available major-league computer algebra programs while you're reading this material. Four of these, any one of which would certainly fill the bill, are Macsyma[2], Maple[3], Mathematica[4], or Axiom[5]. What one needs from such programs are a large number of high level mathematical commands and a built-in programming language. In this book we will for the most part use Maple and Mathematica, and we will also discuss some public domain packages that are available.

2.2 Identities

An identity is a mathematical equation that states that two seemingly different things are in fact the same, at least under certain conditions. So "$2 + 2 = 4$" is an identity, though perhaps not a shocker. So is "$(x + 1)^2 = 1 + 2x + x^2$," which is a more advanced specimen because it has a free parameter "x" in it, and the statement is true for all (real, complex) values of x.

There are beautiful identities in many branches of mathematics. Number theory, for instance, is one of their prime habitats:

$$95800^4 + 217519^4 + 414560^4 = 422481^4$$

$$\sum_{k \backslash n} \mu(k) = \begin{cases} 1, & \text{if } n = 1; \\ 0, & \text{if } n \geq 2, \end{cases}$$

$$\prod_p (1 - p^{-s})^{-1} = \sum_{n \geq 1} \frac{1}{n^s} \qquad (\mathrm{Re}\,(s) > 1),$$

$$\det\left((\gcd(i, j))_{i,j=1}^n\right) = \phi(1)\phi(2) \cdots \phi(n),$$

$$1 + \sum_{m=1}^\infty \frac{x^{m^2}}{(1 - x)(1 - x^2) \cdots (1 - x^m)} = \prod_{m=0}^\infty \frac{1}{(1 - x^{5m+1})(1 - x^{5m+4})}.$$

---

[2] Macsyma is a product of Symbolics, Inc.
[3] Maple is a product of Waterloo Maple Software, Inc.
[4] Mathematica is a product of Wolfram Research, Inc.
[5] Axiom is a product of NAG (Numerical Algorithms Group), Ltd.

Combinatorics is one of the major producers of marvelous identities:

$$\sum_{j=0}^{n} \binom{n}{j}^2 = \binom{2n}{n}, \tag{2.2.1}$$

$$\sum_{i+j+k=n} \binom{i+j}{i}\binom{j+k}{j}\binom{k+i}{k} = \sum_{0 \le j \le n} \binom{2j}{j}, \tag{2.2.2}$$

$$\exp\left\{\sum_{m \ge 1} m^{m-1}\frac{t^m}{m!}\right\} = 1 + \sum_{n \ge 1}(n+1)^{n-1}\frac{t^n}{n!}, \tag{2.2.3}$$

$$\sum_{s=0}^{2m}(-1)^s\binom{2m}{s}^3 = (-1)^m\frac{(3m)!}{(m!)^3}. \tag{2.2.4}$$

Beautiful identities have often stimulated mathematicians to find correspondingly beautiful proofs for them; proofs that have perhaps illuminated the combinatorial or other significance of the equality of the two members, or possibly just dazzled us with their unexpected compactness and elegance. It is a fun activity for people to try to prove identities. We have been accused of taking the fun out of it by developing these computer methods[6] but we hope that we have in fact only moved the fun to a different level.

Here we will not, of course, be able to discuss all kinds of identities. Far from it. We are going to concentrate on one family of identities, called hypergeometric identities, that have been of great interest and importance, and include many of the famous binomial coefficient identities of combinatorics, such as equations (2.2.1), (2.2.2) and (2.2.4) above.

The main purpose of this book is to explain how the discoveries and the proofs of hypergeometric identities have been very largely automated. The book is not primarily about computing; it is the mathematics that underlies the computing that will be the main focus. Automating the discovery and proof of identities is not something that is immediately obvious as soon as you have a large computer. The theoretical developments that have led to the automation make what we believe is a very interesting story, and we would like to tell it to you.

The proof theory of these identities has gone through roughly three phases of evolution.

At first each identity was treated on its own merits. Combinatorial insights proved some, generating functions proved others, special tricks proved many, but unified methods of wide scope were lacking, although many of the special methods were ingenious and quite effective.

---

[6]See How the Grinch stole mathematics [Cipr89].

In the next phase it was recognized that a very large percentage of combinatorial identities, the ones that involve binomial coefficients and factorials and such, were in fact special cases of a few very general hypergeometric identities. The theory of hypergeometric functions was initiated by C. F. Gauss early in the nineteenth century, and in the course of developing that theory some very general identities were found. It was not until 1974, however, that the recognition mentioned above occurred. There was, therefore, a considerable time lag between the development of the "new technology" of hypergeometric identities, and its "application" to binomial coefficient sums of combinatorics.

A similar, but much shorter, time lag took place before the third phase of the proof theory flowered. In the 1940s, the main ideas for the *automated* discovery of recurrence relations for hypergeometric sums were discovered by Sister Mary Celine Fasenmyer (see Chapter 4). It was not until 1982 that it was recognized, by Doron Zeilberger [Zeil82], that these ideas also provided tools for the automated proofs of hypergeometric identities. The essence of what he recognized was that if you want to prove an identity

$$\sum_k \text{summand}(n, k) = \text{answer}(n) :$$

then you can

- FIND A RECURRENCE RELATION THAT IS SATISFIED BY THE SUM ON THE LEFT SIDE.
- SHOW THAT ANSWER(n) SATISFIES THE SAME RECURRENCE, BY SUBSTITUTION.
- CHECK THAT ENOUGH CORRESPONDING INITIAL VALUES OF BOTH SIDES ARE EQUAL TO EACH OTHER.

With that realization the idea of finding recurrence relations that sums satisfy was elevated to the first priority task in the analysis of identities. As the many facets of that realization have been developed, the emergence of powerful high level computer algebra programs for personal computers and workstations has brought the whole chain of ideas to your own desktop. Anyone who has access to such equipment can use the programs of this book, or others that are available, to prove and discover many kinds of identities.

2.3 Human and computer proofs; an example

In this section we are going to take one identity and illustrate the evolution of proof theory by proving it in a few different ways. The identity that we'll use is

$$\sum_k \binom{n}{k}^2 = \binom{2n}{n}. \tag{2.3.1}$$

First we present a purely combinatorial proof. There are $\binom{n}{k}$ ways to choose k letters from among the letters $1, 2, \ldots, n$. There are $\binom{n}{n-k}$ ways to choose $n - k$ letters from among the letters $n + 1, \ldots, 2n$. Hence there are $\binom{n}{k}\binom{n}{n-k} = \binom{n}{k}^2$ ways to make such a pair of choices. But every one of the $\binom{2n}{n}$ ways of choosing n letters from the $2n$ letters $1, 2, \ldots, 2n$ corresponds uniquely to such a pair of choices, for some k.   □

We must pause to remark that that one is a really nice proof. So as we go through this book whose main theme is that computers can prove all of these identities, please note that we will never[7] claim that computerized proofs are *better* than human ones, in any sense. When an elegant proof exists, as in the above example, the computer will be hard put to top it. On the other hand, the contest will be close even here, because the computerized proof that's coming up is rather elegant too, in a different way.

To continue, the pre-computer proof of (2.3.1) that we just gave was combinatorial, or bijective. It found the combinatorial interpretations of both sides of the identity, and showed that they both count the same thing.

Here's another vintage proof of the same identity. The coefficient of x^r in $(1 + x)^{a+b}$ is obviously $\binom{a+b}{r}$. On the other hand, the coefficient of x^r in $(1+x)^a(1+x)^b$ is, just as obviously, $\sum_k \binom{a}{k}\binom{b}{r-k}$, and these two expressions for the same coefficient must be equal. Now take $a = b = r = n$.   □

That was a proof by generating functions, another of the popular tools used by the species *Homo sapiens* for the proof of identities before the computer era.

Next we'll show what a computerized proof of the same identity looks like. We preface it with some remarks about *standardized proofs* and *certificates*.

Suppose we're going to develop machinery for proving some general category of theorems, a category that will have thousands of individual examples. Then it would clearly be nice to have a rather standardized proof outline, one that would work on all of the thousands of examples. Now somehow each example is different. So the proofs have to be a little bit different as we pass from one of the thousands of examples to another. The trick is to get the proofs to be as identical as possible, differing in only some single small detail. That small detail will be called the *certificate*. Since the rest of the proof is standard, and not dependent on the particular example, we will be able to describe the complete proof for a given example just by describing the proof certificate.

In the case of proving binomial coefficient identities, the WZ method is a standardized proof procedure that is almost independent of the particular identity that you're trying to prove. The only thing that changes in the proof, as we go from one

---

[7]Well, hardly ever.

identity to another, is a certain rational function $R(n, k)$ of two variables, n and k. Otherwise, all of the proofs are the same.

So when your computer finds a WZ proof, it doesn't have to recite the whole thing; it needs to describe only the rational function $R(n, k)$ that applies to the particular identity that you are trying to prove. The rest of the proof is standardized. The rational function $R(n, k)$ certifies the proof.

Here is the standardized WZ proof algorithm:

1. Suppose that you wish to prove an identity of the form $\sum_k t(n, k) = \mathrm{rhs}(n)$, and let's assume, for now, that for each n it is true that the summand $t(n, k)$ vanishes for all k outside of some finite interval.

2. Divide through by the right hand side, so the identity that you wish to prove now reads as $\sum_k F(n, k) = 1$, where $F(n, k) = t(n, k)/\mathrm{rhs}(n)$.

3. Let $R(n, k)$ be the rational function that the WZ method provides as the proof of your identity (we'll discuss how to find this function in Chapter 7). Define a new function $G(n, k) = R(n, k)F(n, k)$.

4. You will now observe that the equation

$$F(n + 1, k) - F(n, k) = G(n, k + 1) - G(n, k)$$

is true. Sum that equation over all integers k, and note that the right side telescopes to 0. The result is that

$$\sum_k F(n + 1, k) = \sum_k F(n, k),$$

hence we have shown that $\sum_k F(n, k)$ is independent of n, i.e., is constant.

5. Verify that the constant is 1 by checking that $\sum_k F(0, k) = 1$.   □

The rational function $R(n, k)$ is the key that turns the lock. The lock is the proof outlined above. If you want to prove an identity, and you have the key, then just put it into the lock and watch the proof come out.

We're going to illustrate the method now with a few examples.

Example 2.3.1. First let's try the venerable identity $\sum_k \binom{n}{k} = 2^n$. The key to the lock, in this case, is the rational function $R(n, k) = k/(2(k - n - 1))$. Please remember that if you want to know *how to find* the key, for a given identity, you'll have to wait at least until page 126. For now we're going to focus on how to use the key, rather than on how to find it.

We'll follow the standardized proof through, step by step.

In Step 1, our term $t(n, k)$ is $\binom{n}{k}$, and the right hand side is $\mathrm{rhs}(n) = 2^n$.

For Step 2, we divide through by 2^n and find that the standardized summand is $F(n, k) = \binom{n}{k} 2^{-n}$, and we now want to prove that $\sum_k F(n, k) = 1$, for this F.

In Step 3 we use the key. We take our rational function $R(n, k) = k/(2(k - n - 1))$, and we define a new function

$$G(n, k) = R(n, k) F(n, k) = \frac{k}{2(k - n - 1)} \binom{n}{k} 2^{-n}$$

$$= -\frac{k n! \, 2^{-n}}{2(n + 1 - k) k! \, (n - k)!} = -\binom{n}{k - 1} 2^{-n-1}.$$

Step 4 informs us that we will have now the equation $F(n + 1, k) - F(n, k) = G(n, k + 1) - G(n, k)$. Let's see if that is so. In other words, is it true that

$$\binom{n + 1}{k} 2^{-n-1} - \binom{n}{k} 2^{-n} = -\binom{n}{k} 2^{-n-1} + \binom{n}{k - 1} 2^{-n-1}?$$

Well, at this point we have arrived at a situation that will be referred to throughout this book as a "routinely verifiable" identity. That phrase means roughly that your pet chimpanzee could check out the equation. More precisely it means this. First cancel out all factors that look like c^n or c^k (in this case, a factor of 2^{-n}) that can be cancelled. Then replace every binomial coefficient in sight by the quotient of factorials that it represents. Finally, cancel out all of the factorials by suitable divisions, leaving only a polynomial identity that involves n and k. After a few more strokes of the pen, or keys on the keyboard, this identity will reduce to the indisputable form $0 = 0$, and you'll be finished with the "routine verification."

In this case, after multiplying through by 2^n, and replacing all of the binomial coefficients by their factorial forms, we obtain

$$\frac{(n + 1)!}{2k! \, (n + 1 - k)!} - \frac{n!}{k! \, (n - k)!} = -\frac{n!}{2k! \, (n - k)!} + \frac{n!}{2(k - 1)! \, (n - k + 1)!}$$

as the equation that is to be "routinely verified." To clear out all of the factorials we multiply through by $k! \, (n + 1 - k)!/n!$, and get

$$\frac{n + 1}{2} - (n + 1 - k) = -\frac{n + 1 - k}{2} + \frac{k}{2},$$

which is really trivial.

In Step 5 of the standardized WZ algorithm we must check that $\sum_k F(0,k) = 1$. But

$$F(0,k) = \binom{0}{k} = \begin{cases} 1, & \text{if } k = 0; \\ 0, & \text{otherwise,} \end{cases}$$

and we're all finished (that's also the last time we'll do a routine verification in full). □

Example 2.3.2.

In an article in the American Mathematical Monthly **101** (1994), p. 356, it was necessary to prove that $\sum_k F(n,k) = 1$ for all n, where

$$F(n,k) = \frac{(n-i)!\,(n-j)!\,(i-1)!\,(j-1)!}{(n-1)!\,(k-1)!\,(n-i-j+k)!\,(i-k)!\,(j-k)!}. \qquad (2.3.2)$$

The complete proof is given by the rational function $R(n,k) = (k-1)/n$ (it is noteworthy that, in this example, $R(n,k)$ does not depend on i or j). □

2.4   A Mathematica session

For our next example of the use of the WZ proof algorithm we'll take some of the pain out by using Mathematica to do the routine algebra.

To begin, let's try to simplify some expressions that contain factorials. If we type in

$$\text{In}[1] := (n+1)!/n!$$

then what we get back is

$$\text{Out}[1] = \frac{(1+n)!}{n!},$$

which doesn't help too much. On the other hand if we enter

$$\text{In}[2] := \text{Simplify}[(n+1)!/n!]$$

then we also get back

$$\text{Out}[2] = \frac{(1+n)!}{n!},$$

so we must be doing something wrong. Well, it turns out that if you would really like to simplify ratios of factorials then the thing to do is to read in the package RSolve, because in that package there lives a command FactorialSimplify, which does the simplification that you would like to see.

So let's start over, this time with

$$\text{In}[1] := << \texttt{DiscreteMath`RSolve`}.$$

Next we ask for

$$\text{In}[2] := \texttt{FactorialSimplify}[(n + 1)!/n!]$$

and we get

$$\text{Out}[2] = 1 + \mathbf{n},$$

which is what we wanted.

Let's now verify the WZ proof of the identity $\sum_k \binom{n}{k}^2 = \binom{2n}{n}$, of (2.3.1). Our standardized summand, obtained by dividing the original identity by its right hand side, is $F(n, k) = n!^4/(k!^2(n - k)!^2(2n)!)$. The rational function certificate (the key to the lock) for this identity is

$$R(n, k) = -\frac{k^2(3n + 3 - 2k)}{2(n + 1 - k)^2(2n + 1)}.$$

So we ask Mathematica to create the function $G(n, k) = R(n, k)F(n, k)$. To do this we first define R,

```
In[3]:= r[n_ ,k_ ] := -k^2 (3n+3-2k)/(2(n+1-k)^2 (2n+1)),
```

and then we define the pair (F, G) of functions that occur in the WZ method by typing

```
In[4]:= f[n_ ,k_ ]:=n!^4/(k!^2 (n-k)!^2 (2n)!)
In[5]:= g[n_ ,k_ ]:=r[n,k] f[n,k].
```

To do the routine verification, you now need only ask for

```
In[6]:= FactorialSimplify[f[n+1,k]-f[n,k]-g[n,k+1]+g[n,k]],
```

and after a few moments of reflection, you will be rewarded with

```
Out[6]=  0
```

which is the name of the game.   □

2.5   A Maple session

Now we're going to try the same thing in Maple. First we try to learn to simplify factorial ratios, so we hopefully type $(n + 1)!/n!$;, and the system responds by giving us back our input unaltered. So it needs to be coaxed. A good way to coax it is with

$$\text{expand}((n + 1)!/n!);$$

and we're rewarded with the $n + 1$ that we were looking for. So Maple's

$$\text{expand}();$$

command is the way to simplify factorial expressions (in some versions of Maple this command does not work properly on quotients of products of factorials in which the factors are raised to powers).

Now let's tell Maple the rational function certificate $R(n, k)$,

$$r := (n, k)- > -k\hat{\ }2 * (3 * n + 3 - 2 * k)/(2 * (n + 1 - k)\hat{\ }2 * (2 * n + 1));$$

and then we input our standardized summand $F(n, k)$ as

$$f := (n, k)- > n!\hat{\ }4/(k!\hat{\ }2 * (n - k)!\hat{\ }2 * (2 * n)!);$$

We ask Maple to define the function $G(n, k) = R(n, k)F(n, k)$,

$$g := (n, k)- > r(n, k) * f(n, k);$$

Now there's nothing to do but see if the basic WZ equation is satisfied. It is best not to ask just for

$$f(n + 1, k) - f(n, k) - g(n, k + 1) + g(n, k),$$

and hope that it will vanish, because a large distributed expression will result. The best approach seems to be to divide the whole expression by $f(n, k)$, then expand it to get rid of all of the factorials , and then simplify it, in order to collect terms. So the recommended command to Maple would be

$$\text{simplify}(\text{expand}((f(n + 1, k) - f(n, k) - g(n, k + 1) + g(n, k))/f(n, k)));$$

which would return the desired output of 0.

2.6   Where we are and what happens next

We have so far discussed the following two pairs, each consisting of an identity and its WZ proof certificate:

$$\sum_k \binom{n}{k} = 2^n, \qquad R(n,k) = \frac{k}{2(k-n-1)};$$

$$\sum_k \binom{n}{k}^2 = \binom{2n}{n}, \qquad R(n,k) = -\frac{k^2(3n+3-2k)}{2(n+1-k)^2(2n+1)}.$$

So what we can expect from computer methods are short, even one-line, proofs of combinatorial identities, in standardized format, as well as finding the right hand side if it is unknown. Human beings might have a great deal of trouble in finding one of these proofs, but the verification procedure, as we have seen, is perfectly civilized, and involves only a medium amount of human labor.

In Chapter 3 we will meet the hypergeometric database. There we will learn how to take identities that involve binomial coefficients and factorials and write them in standard hypergeometric form. We will see that this gives us access to a database of theorems and identities, and we will learn how to interrogate that database. We will also see some of its limitations.

Following this are five consecutive chapters that deal with the fundamental algorithms of the subject of computer proofs of identities.

Chapter 4 describes the original algorithm of Sister Mary Celine Fasenmyer. She developed, in her doctoral dissertation of 1945, the first computerizable method for finding recurrence relations that are satisfied by sums. We also prove the validity of her algorithm here, since that fact underlies the later developments.

Chapter 5 is about the fundamental algorithm of Gosper, which is to summation as finding antiderivatives is to integration. This algorithm allows us to do *indefinite* hypergeometric sums in simple closed form, or it furnishes a proof of impossibility if, in a given case, that cannot be done. Beyond its obvious use in doing indefinite sums, it has several nonobvious uses in executing the WZ method, in finding recurrences for definite sums, and even for finding the right hand side of a definite sum whose evaluation we are seeking.

Chapter 6 deals with Zeilberger's algorithm ("creative telescoping"). Again, this is an algorithm that finds recurrence relations that are satisfied by sums. It is in most cases much faster than the method of Sister Celine, and it has made possible a whole generation of computerized proofs of identities that were formerly inaccessible to these ideas. It is the cornerstone of the methods that we present for finding out if a given combinatorial sum can be simplified, and *it is guaranteed to work every time*.

Chapter 7 contains a complete discussion of the "WZ phenomenon." This method, which we have already previewed here, just to get you interested, provides by far the most compact certifications of combinatorial identities, though it does not exist in the full generality of the methods of Chapters 4 and 6. When it does exist, however, which is very often, it gives us the unique opportunity to find new identities, as well as to prove old ones. We will see here how to do that, and give some examples of the treasures that can be found in this way.

In Chapter 8 we deal with the question of solving linear recurrences with polynomial coefficients. In all of the examples earlier in the book, the computer analysis of an identity will produce just such a recurrence that the sum satisfies. If we want to prove that a certain right hand side is the correct one, then we just check that the claimed right hand side satisfies the same recurrence and we check a few initial conditions.

But suppose we don't know the right hand side. Then we have a recurrence with polynomial coefficients that our sum satisfies, and we want to know if it has, in a certain sense, a simple solution. If the recurrence happens to be of first order, we're finished. But what if it isn't? Then we need to know how to recognize when a higher order recurrence has simple solutions of a certain form, and when it does not. The fundamental algorithm of this subject, due to Petkovšek, is in Chapter 8 (see page 156).

2.7 Exercises

1. Let $f(n) = (3n+1)!\,(2n-5)!/(n+2)!^2$. Use a computer algebra program to exhibit

$$\sum_{k=0}^{3} \frac{f(n-k)}{f(n)}$$

explicitly as a quotient of two polynomials in n.

2. Use a computer algebra program to check the following pairs. Each pair consists of an identity and its WZ proof certificate $R(n,k)$:

$$\sum_k (-1)^k \binom{n}{k} \frac{x}{k+x} = \frac{1}{\binom{x+n}{n}}, \qquad \frac{k(k+x)}{(n+1)(k-n-1)}$$

$$\sum_k \binom{n}{k}\binom{x}{k+r} = \binom{n+x}{n+r}, \qquad \frac{k(k+r)}{(n+x+1)(k-n-1)}$$

$$\sum_k \binom{x+1}{2k+1}\binom{x-2k}{n-k} 2^{2k+1} = \binom{2x+2}{2n+1}, \qquad \frac{k(2k+1)(x-2n-1)}{(k-n-1)(2n-2x-1)(n-x)}$$

$$(1 - 2n) \sum_k (-1)^k \frac{\binom{n}{k} 4^k}{\binom{2k}{k}} = 1, \qquad \frac{k(2k-1)}{(2n-1)(k-n-1)}.$$

3. For each of the four parts of Problem 2 above, write out the complete proof of the identity, using the full text of the standardized WZ proof together with the appropriate rational function certificate.

4. For each of the parts of Problem 2 above, say exactly what the standardized summand $F(n, k)$ is, and in each case evaluate

$$\lim_{k \to \infty} F(n, k) \qquad \text{and} \qquad \lim_{n \to \infty} F(n, k).$$

5. Write a procedure, in your favorite programming language, whose input will be the summand $t(n, k)$, and the right hand side rhs(n), of a claimed identity $\sum_k t(n, k) = \text{rhs}(n)$, as well as a claimed WZ proof certificate $R(n, k)$. Output of the procedure should be "The claimed identity has been verified," or "Error; the claimed identity has not been proved," depending on how the verification procedure turns out. Test your program on the examples in Problem 2 above. Be sure to check the initial conditions as well as the WZ equation.

Chapter 3

The Hypergeometric Database

3.1 Introduction

In this book, which is primarily about sums, *hypergeometric* sums occupy center stage. Roughly (see the formal definition below), a hypergeometric sum is one in which the summand involves only factorials, polynomials, and exponential functions of the summation variable. This class includes multitudes of sums that contain binomial coefficients and factorials, including virtually all of the familiar ones that have been summed in closed form.[1]

The fact is that many hypergeometric sums can be expressed in simple closed form, and many others can be revealed to be equal to some other, seemingly different, hypergeometric sum. Whenever this happens we have an identity. Typically, when we look at such an identity we will see an equation that has on the left hand side a sum in which the summand contains a number of factorials, binomial coefficients, etc., and has on the right a considerably simpler function that is equal to the sum on the left. In Chapter 2 we saw a few examples of such identities. If you would like to see what a complicated identity looks like, try this one, which holds for integer n:

$$\sum_{r,s}(-1)^{n+r+s}\binom{n}{r}\binom{n}{s}\binom{n+r}{r}\binom{n+s}{s}\binom{2n-r-s}{n} = \sum_k\binom{n}{k}^4.$$

In dealing with these sums it may be important to have a standard notation and classification. There is such a wealth of information available now that it is important to have systematic ways of searching the literature for information that may help us to deal with a particular sum.

---

[1]See page 145 for a precise definition of "closed form."

So our main task in this chapter will be to show how a given sum is described by using standardized hypergeometric notation. Once we have that in hand, it will be much easier to consult databases of known information about such sums. An entry in such a database is a statement to the effect that a certain hypergeometric series is equal to a certain much simpler expression, for all values of the various free parameters that appear, or at least for all values in a suitably restricted range.

We must emphasize that the main thrust of this book is *away* from this approach, to look instead at an alternative to such database lookups. We will develop computerized methods of such generality and scope that instead of attempting to look up a sum in such a database, which is a process that is far from algorithmic, and which has no theorem that guarantees success under general conditions, it will often be preferable to ask the computer to prove the identity directly or to find out if a simple evaluation of it exists. Nevertheless, hypergeometric function theory is the context in which this activity resides, and the language of that theory, and its main theorems, are important in all of these applications.

3.2 Hypergeometric series

A *geometric* series $\sum_{k \geq 0} t_k$ is one in which the ratio of every two consecutive terms is constant, i.e., t_{k+1}/t_k is a constant function of the summation index k. The kth term of a geometric series is of the form cx^k where c and x are constants, i.e., are independent of the summation index k. Therefore a general *geometric* series looks like

$$\sum_{k \geq 0} cx^k.$$

A *hypergeometric* series $\sum_{k \geq 0} t_k$ is one in which $t_0 = 1$ and the ratio of two consecutive terms is a *rational function* of the summation index k, i.e., in which

$$\frac{t_{k+1}}{t_k} = \frac{P(k)}{Q(k)},$$

where P and Q are polynomials in k. In this case we will call the terms *hypergeometric terms*. Examples of such hypergeometric terms are $t_k = x^k$, or $k!$, or $(2k+7)!/(k-3)!$, or $(k^2-1)(3k+1)!/((k+3)!\,(2k+7))$.

Hypergeometric series are very important in mathematics. Many of the familiar functions of analysis are hypergeometric. These include the exponential, logarithmic, trigonometric, binomial, and Bessel functions, along with the classical orthogonal polynomial sequences of Legendre, Chebyshev, Laguerre, Hermite, etc.

It is important to recognize when a given series is hypergeometric, if it is, because the general theory of hypergeometric functions is very powerful, and we may gain a lot of insight into a function that concerns us by first recognizing that it is hypergeometric, then identifying precisely which hypergeometric function it is, and finally by using known results about such functions.

In the ratio of consecutive terms, $P(k)/Q(k)$, let us imagine that the polynomials P and Q have been completely factored, in the form

$$\frac{t_{k+1}}{t_k} \overset{def}{=} \frac{P(k)}{Q(k)} = \frac{(k+a_1)(k+a_2)\cdots(k+a_p)}{(k+b_1)(k+b_2)\cdots(k+b_q)(k+1)} x, \qquad (3.2.1)$$

where x is a constant. If we normalize by taking $t_0 = 1$, then we denote the hypergeometric series whose terms are the t_k's, i.e., the series $\sum_{k\geq 0} t_k x^k$, by

$$
{}_pF_q\!\begin{bmatrix} a_1 & a_2 & \cdots & a_p \\ b_1 & b_2 & \cdots & b_q \end{bmatrix}; x \Big].
$$

The a's and the b's are called, respectively, the *upper* and the *lower* parameters of the series. The b's are not permitted to be nonpositive integers or the series will obviously not make sense.

To put it another way, the hypergeometric series

$$
{}_pF_q\!\begin{bmatrix} a_1 & a_2 & \cdots & a_p \\ b_1 & b_2 & \cdots & b_q \end{bmatrix}; x \Big]
$$

is the series whose initial term is 1, and in which the ratio of the $(k+1)^{\text{st}}$ term to the kth is given by (3.2.1) above, for all $k \geq 0$.

The appearance of the factor $(k+1)$ in the denominator of (3.2.1) needs a few words of explanation. Why, one might ask, does it have to be segregated that way, when it might just as well have been absorbed as one of the factors $(k+b_i)$? The answer is that there's no reason except that it has always been done that way, and far be it from us to try to reverse the course of the Nile. If there is not a factor of $(k+1)$ in the denominator of your term ratio, i.e., if there is no factor of $k!$ in the denominator of your term t_k, just put it in, and compensate for having done so by putting an extra factor in the numerator.

3.3 How to identify a series as hypergeometric

Many of the famous functions of classical analysis have hypergeometric series expansions. In the exponential series, $e^x = \sum_{k\geq 0} x^k/k!$, the initial term is 1, and the ratio of the $(k+1)^{\text{st}}$ term to the kth is $x/(k+1)$, which is certainly of the form of (3.2.1). So $e^x = {}_0F_0\!\begin{bmatrix} - \\ - \end{bmatrix}; x \Big]$.

What we want to do now is to show how one may identify a given hyperge-ometric series as a particular $_pF_q[\cdots]$. This process is at the heart of using the hypergeometric database. If we have a hypergeometric series that interests us for some reason, we might wonder what is known about it. Is it possible to sum the series in simple form? Is it possible to transform the series into another form that is easier to work with? Is some result that we have just discovered about this series really new or is it well known? These questions can often be answered by consulting the extensive literature on hypergeometric series. But the first step is to rewrite the series that interests us in the standard $_pF_q[\cdots]$ form, because the literature is cast in those terms.

Example 3.3.1. If the kth term of the series is $t_k = 2^k/k!^2$, then we have $t_{k+1}/t_k = 2/(k+1)^2$ which is in the form of (3.2.1) with $p = 0$, $q = 1$, and $x = 2$. Consequently, since $t_0 = 1$, the given series is

$$\sum_{k\geq0}\frac{2^k}{k!^2} = {_0F_1}\!\left[\begin{matrix}-\\1\end{matrix};2\right].$$

$\square$

The hypergeometric series lookup algorithm

1. Given a series $\sum_k t_k$. Shift the summation index k so that the sum starts at $k = 0$ with a nonzero term. Extract the term corresponding to $k = 0$ as a common factor so that the first term of the sum will be 1.

2. Simplify the ratio t_{k+1}/t_k to bring it into the form $P(k)/Q(k)$, where P, Q are polynomials. If this cannot be done, the series is not hypergeometric.

3. Completely factor the polynomials P and Q into linear factors, and write the term ratio in the form

$$\frac{P(k)}{Q(k)} = \frac{(k + a_1)(k + a_2)\cdots(k + a_p)}{(k + b_1)(k + b_2)\cdots(k + b_q)(k + 1)}x$$

If the factor $k+1$ in the denominator wasn't there, put it in, and compensate by inserting an extra factor of $k + 1$ in the numerator. Notice that all of the coefficients of k, in numerator and denominator, are $+1$. Whatever numerical factors are needed to achieve this are absorbed into the factor x.

4. You have now identified the input series. It is (the common factor that you extracted in step 1 above, multiplied by) the hypergeometric series

$$_pF_q\!\left[\begin{matrix}a_1 & a_2 & \cdots & a_p\\b_1 & b_2 & \cdots & b_q\end{matrix};x\right]. \square$$

Example 3.3.2. Consider the series $\sum_k t_k$ where $t_k = 1/((2k + 1)(2k + 3)!)$.

To identify this series, note that the smallest value of k for which the term t_k is nonzero is the term with $k = -1$. Hence we begin by shifting the origin of the sum as follows:

$$\sum_{k \geq -1} \frac{1}{(2k + 1)(2k + 3)!} = \sum_{k \geq 0} \frac{1}{(2k - 1)(2k + 1)!}.$$

The ratio of two consecutive terms is

$$\frac{t_{k+1}}{t_k} = \frac{(k - \frac{1}{2})}{(k + \frac{1}{2})(k + \frac{3}{2})(k + 1)} \frac{1}{4}. \tag{3.3.1}$$

Hence our given series is identified as

$$\sum_k \frac{1}{(2k + 1)(2k + 3)!} = - {}_1F_2\begin{bmatrix} -\frac{1}{2} \\ \frac{1}{2} & \frac{3}{2} \end{bmatrix} ; \frac{1}{4} \end{bmatrix}.$$

□

Example 3.3.3. Suppose we define the symbol

$$[x, d]_n = \begin{cases} \prod_{j=0}^{n-1}(x - jd), & \text{if } n > 0; \\ 1, & \text{if } n = 0. \end{cases}$$

Now consider the series

$$\sum_k \binom{n}{k} [x, d]_k [y, d]_{n-k}.$$

Is this a hypergeometric series, and if so which one is it?

The term ratio is

$$\begin{aligned}
\frac{t_{k+1}}{t_k} &= \frac{\binom{n}{k+1}[x, d]_{k+1}[y, d]_{n-k-1}}{\binom{n}{k}[x, d]_k[y, d]_{n-k}} \\
&= \frac{(n - k)\prod_{j=0}^{k}(x - jd)\prod_{j=0}^{n-k-2}(y - jd)}{(k + 1)\prod_{j=0}^{k-1}(x - jd)\prod_{j=0}^{n-k-1}(y - jd)} \\
&= \frac{(n - k)(x - kd)}{(k + 1)(y - (n - k - 1)d)} = \frac{(k - n)(k - \frac{x}{d})}{(k + (\frac{y}{d} - n + 1))(k + 1)}.
\end{aligned}$$

This is exactly in the standard form (3.2.1), so we have identified our series as the hypergeometric series

$$([y, d]_n) \, {}_2F_1\begin{bmatrix} -n & -\frac{x}{d} \\ \frac{y}{d} - n + 1 \end{bmatrix} 1 \end{bmatrix}.$$

□

Example 3.3.4. Suppose we are wondering if the sum

$$\sum_{k=0}^{n} \binom{n}{k} \frac{(-1)^k}{k!}$$

can be evaluated in some simple form. A first step might be to identify it as a hypergeometric series.[2] The next step would then be to look up that hypergeometric series in the database to see if anything is known about it. Let's do the first step here.

The term ratio is

$$\frac{t_{k+1}}{t_k} = \frac{\binom{n}{k+1} \frac{(-1)^{k+1}}{(k+1)!}}{\binom{n}{k} \frac{(-1)^k}{k!}}$$
$$= \frac{k-n}{(k+1)^2},$$

and $t_0 = 1$. Hence by (3.2.1) our unknown sum is revealed to be a

$$_1F_1 \left[\begin{matrix} -n \\ 1 \end{matrix} ; 1 \right].$$

$\square$

Example 3.3.5. Is the Bessel function

$$J_p(x) = \sum_{k=0}^{\infty} \frac{(-1)^k (\frac{x}{2})^{2k+p}}{k! \, (k+p)!}$$

a hypergeometric function? The ratio of consecutive terms is

$$\frac{t_{k+1}}{t_k} = \frac{(-1)^{k+1}(\frac{x}{2})^{2k+2+p} k! \, (k+p)!}{(k+1)! \, (k+p+1)!(-1)^k(\frac{x}{2})^{2k+p}}$$
$$= \frac{-(\frac{x^2}{4})}{(k+1)(k+p+1)}.$$

Here we must take note of the fact that $t_0 \neq 1$, whereas the standardized hypergeometric series begins with a term equal to 1. Our conclusion is that the Bessel function is indeed hypergeometric, and it is in fact

$$J_p(x) = \frac{(\frac{x}{2})^p}{p!} \; _0F_1 \left[\begin{matrix} - \\ p+1 \end{matrix} ; -\frac{x^2}{4} \right].$$

$\square$

---

[2]We hope to convince you that a better first step is to reach for your computer!

We will use the notation

$$(a)_n \overset{def}{=} \begin{cases} a(a+1)(a+2)\cdots(a+n-1), & \text{if } n \geq 1; \\ 1, & \text{if } n = 0. \end{cases}$$

for the *rising factorial* function.

In terms of the rising factorial function, here is what the general hypergeometric series looks like:

$$_pF_q\begin{bmatrix} a_1 & a_2 & \dots & a_p \\ b_1 & b_2 & \dots & b_q \end{bmatrix} = \sum_{k\geq 0} \frac{(a_1)_k(a_2)_k\cdots(a_p)_k}{(b_1)_k(b_2)_k\cdots(b_q)_k} \frac{z^k}{k!}. \qquad (3.3.2)$$

The series is well defined as long as the *lower parameters* $b_1, b_2, \ldots, b_q$ are not negative integers or zero. The series *terminates* automatically if any of the *upper parameters* $a_1, a_2, \ldots, a_p$ is a nonpositive integer, otherwise it is *nonterminating*, i.e., it is an infinite series. If the series is well defined and nonterminating, then questions of convergence or divergence become relevant. In this book we will be concerned for the most part with terminating series.

3.4 Software that identifies hypergeometric series

The act of taking a series and finding out exactly which $_pF_q[\ldots]$ it is can be fairly tedious. Computers can help even with this humble task. In this section we'll discuss the use of Mathematica, Maple, and a special purpose package *Hyp* that was developed by C. Krattenthaler.

First, Mathematica has a limited capability for transforming sums that are given in customary summation form into standard hypergeometric form. This capability resides in the package `Algebra'SymbolicSum'`. So we first read in the package, with `<<Algebra'SymbolicSum'`. Now we try it with

$$\text{Sum[Binomial[n, k]\textasciicircum 2, \{k, 0, n\}],}$$

hoping to find out which hypergeometric sum this is. But Mathematica is too smart for us. It knows how to evaluate the sum in simple form, and so it proudly replies

$$\frac{(2n)!}{n!\textasciicircum 2},$$

which in this instance is more than we were asking for. We can force Mathematica to identify our sum as a $_pF_q$ only when it does not know how to express it in simple form. Here a good tactic would be to insert an extra x^k into the sum, hope that

Mathematica does not know any simple form for that one, and then put $x = 1$ in the answer. So we hopefully type

$$\texttt{Sum[Binomial[n,k]\^{}2 x\^{}k, \{k,0,n\}],}$$

and sure enough it responds with

$$\texttt{Hypergeometric2F1[-n, -n, 1, x],}$$

and now we can let $x = 1$ to learn that our original sum was a $_2F_1\left[\begin{matrix} -n, -n \\ 1 \end{matrix}; 1\right]$.

Next let's try the sum in Example 3.3.4 above. If we enter

$$\texttt{Sum[Binomial[n,k] (-1)\^{}k/k!, \{k,0,n\}],}$$

we find that Mathematica is very well trained indeed, since it gives

$$\texttt{LaguerreL[n, 0, 1]}$$

which means that it recognizes our sum as a Laguerre polynomial! The trick of inserting x^k won't change this behavior, so there isn't any way to adapt this routine to the present example.

In Mathematica, when we cannot get the $\texttt{SymbolicSum}$ package to identify a sum for us, we can change our strategy slightly, and use Mathematica to help us identify the sum "by hand." In the above case we would first define the general kth term of our sum,

$$\texttt{t[k\_] := (-1)\^{}k n!/(k!\^{}2 (n-k)!)}$$

and then ask for[3] the term ratio,

$$\texttt{FactorialSimplify[t[k+1]/t[k]].}$$

We would obtain the term ratio in the nicely factored form

$$\frac{k - n}{(k + 1)^2}.$$

We would then compare this with (3.2.1) to find that the input series was, as in Example 3.3.4,

$$_1F_1\left[\begin{matrix} -n \\ 1 \end{matrix}; 1\right].$$

---

[3]Read in $\texttt{DiscreteMath`RSolve`}$ before attempting to $\texttt{FactorialSimplify}$ something.

To finish on a positive note, we'll ask Mathematica to identify quite a tricky sum for us, by entering

$$\text{Sum}[(-1)^{\wedge}\text{kBinomial}[r - s - k, k]\text{Binomial}[r - 2k, n - k]/(r - n - k + 1), \{k, 0, n\}].$$
$$(3.4.1)$$

This time it answers us with

$$\frac{\binom{r}{n}}{(r - n + 1)} \; {}_4F_3\left[\begin{matrix} -n, n - r - 1, (s - r)/2, (s - r + 1)/2 \\ (1 - r)/2, -r/2, s - r \end{matrix}; 1\right], \qquad (3.4.2)$$

which is extremely helpful.

Next let's try a session with Maple. The capability in Maple to identify a series as a $_pF_q[\cdots]$ rests with the function **convert/hypergeom**. To identify the sum of the cubes of the binomial coefficients of order n as a hypergeometric series, enter

$$\text{convert}(\text{sum}(\text{binomial}(n, k)^{\wedge}3, k = 0..\text{infinity}), \text{hypergeom});$$

and Maple will answer you with

$$\text{hypergeom}([-n, -n, -n], [1, 1], -1),$$

i.e., with

$$_3F_2\left[\begin{matrix} -n, -n, -n \\ 1, 1 \end{matrix}; -1\right].$$

The "tricky" sum in (3.4.1) can be handled by first defining the summand
$$\text{f:=k->(-1)^{\wedge}k*binomial(r-s-k,k)*binomial(r-2*k,n-k)/(r-n-k+1)};$$

and then making the request

$$\text{convert}(\text{sum}(f(k), k = 0..\text{infinity}), \text{hypergeom});$$

Maple will rise to the occasion by giving the answer as in (3.4.2) above.

Finally we illustrate the use of the package *Hyp*. This package, whose purpose is to facilitate the manipulation of hypergeometric series, can be obtained at no cost by anonymous ftp from `pap.univie.ac.at`, at the University of Vienna, in Austria. It is written in Mathematica source code and must be used in conjunction with Mathematica.

To use it to identify a hypergeometric series involves the following steps. First enter the sum that interests you using the usual Sum construct. Give the expression a name, say `mysum`. Then execute `mysum=mysum//.SumF`, and you will, or should, be looking at the hypergeometric designation of your sum as output.

As an example, take the Laguerre polynomial that we tried in Mathematica. We enter

$$\texttt{mysum} = \texttt{Sum[Binomial[n,k] (-1)\^k/k!, \{k,0,Infinity\}]},$$

and then `mysum=mysum//.SumF`. The output will be the desired hypergeometric form $_1F_1\begin{bmatrix} -n \\ 1 \end{bmatrix}; 1\end{bmatrix}$.

3.5   Some entries in the hypergeometric database

The hypergeometric database can be thought of as the collection of all known hypergeometric identities. The following are some of the most useful database entries. We will not prove any of them just now because *all* of their proofs will follow instantly from the computer certification methods that we will develop in Chapters 4–7.

On the right hand sides of these identities you will find any of three different widely used notations: rising factorial, factorial, and gamma function. The gamma function, $\Gamma(z)$, is defined by

$$\Gamma(z) = \int_0^\infty t^{z-1} e^{-t} dt,$$

if $\mathrm{Re}\,(z) > 0$, and elsewhere by analytic continuation. If z is a nonnegative integer then $\Gamma(z+1) = z!$. Hence the gamma function extends the definition of $n!$ to values of n other than the nonnegative integers. In fact $n!$ is thereby defined for all complex numbers n other than the negative integers. Some of the relationships between these three notations are

$$\Gamma(n+1) = n! = (1)_n,$$

$$(a)_n = a(a+1)\cdots(a+n-1) = \frac{(a+n-1)!}{(a-1)!} = \frac{\Gamma(n+a)}{\Gamma(a)}.$$

(I) Gauss's $_2F_1$ identity. If b is a nonpositive integer or $c - a - b$ has positive real part, then

$$_2F_1\begin{bmatrix} a & b \\ c \end{bmatrix}; 1\end{bmatrix} = \frac{\Gamma(c-a-b)\Gamma(c)}{\Gamma(c-a)\Gamma(c-b)}.$$

(II) Kummer's $_2F_1$ identity. If $a - b + c = 1$, then

$$_2F_1\begin{bmatrix} a & b \\ c \end{bmatrix}; -1\end{bmatrix} = \frac{\Gamma(\frac{b}{2}+1)\Gamma(b-a+1)}{\Gamma(b+1)\Gamma(\frac{b}{2}-a+1)}.$$

If b is a negative integer, then this identity should be used in the form

$$_2F_1\begin{bmatrix} a & b \\ & c \end{bmatrix}; -1\end{bmatrix} = 2\cos\left(\frac{\pi b}{2}\right)\frac{\Gamma(|b|)\Gamma(b-a+1)}{\Gamma(\frac{|b|}{2})\Gamma(\frac{b}{2}-a+1)},$$

which follows from the first form by using the reflection formula

$$\Gamma(z)\Gamma(1-z) = \frac{\pi}{\sin\pi z} \qquad (3.5.1)$$

for the Γ-function and taking the limit as b approaches a negative integer.

(III) Saalschütz's $_3F_2$ identity. If $d+e = a+b+c+1$ and c is a negative integer, then

$$_3F_2\begin{bmatrix} a & b & c \\ d & e \end{bmatrix}; 1\end{bmatrix} = \frac{(d-a)_{|c|}(d-b)_{|c|}}{(d)_{|c|}(d-a-b)_{|c|}}.$$

(IV) Dixon's identity. In prettier and easier-to-remember form this identity reads as

$$\sum_k (-1)^k \binom{a+b}{a+k}\binom{a+c}{c+k}\binom{b+c}{b+k} = \frac{(a+b+c)!}{a!\,b!\,c!}.$$

Translated into formal hypergeometric language, it becomes the statement that, if $1 + \frac{a}{2} - b - c$ has positive real part, and if $d = a - b + 1$ and $e = a - c + 1$, then

$$_3F_2\begin{bmatrix} a & b & c \\ d & e \end{bmatrix}; 1\end{bmatrix} = \frac{(\frac{a}{2})!\,(a-b)!\,(a-c)!\,(\frac{a}{2}-b-c)!}{a!\,(\frac{a}{2}-b)!\,(\frac{a}{2}-c)!\,(a-b-c)!}.$$

(V) Clausen's $_4F_3$ identity. If d is a nonpositive integer and $a + b + c - d = \frac{1}{2}$, and $e = a + b + \frac{1}{2}$, and $a + f = d + 1 = b + g$, then

$$_4F_3\begin{bmatrix} a & b & c & d \\ e & f & g \end{bmatrix}; 1\end{bmatrix} = \frac{(2a)_{|d|}(a+b)_{|d|}(2b)_{|d|}}{(2a+2b)_{|d|}(a)_{|d|}(b)_{|d|}}.$$

(VI) Dougall's $_7F_6$ identity. If $n + 2a_1 + 1 = a_2 + a_3 + a_4 + a_5$ and

$$a_6 = 1 + \frac{a_1}{2}; a_7 = -n; \text{ and } b_i = 1 + a_1 - a_{i+1} \quad (i = 1, \ldots, 6),$$

then

$$_7F_6\begin{bmatrix} a_1 & a_2 & a_3 & a_4 & a_5 & a_6 & a_7 \\ b_1 & b_2 & b_3 & b_4 & b_5 & b_6 \end{bmatrix}; 1\end{bmatrix}$$

$$= \frac{(a_1+1)_n(a_1-a_2-a_3+1)_n(a_1-a_2-a_4+1)_n(a_1-a_3-a_4+1)_n}{(a_1-a_2+1)_n(a_1-a_3+1)_n(a_1-a_4+1)_n(a_1-a_2-a_3-a_4+1)_n}.$$

More identities

$$_1F_0\!\left[\begin{matrix} a \\ - \end{matrix}; z\right] = \frac{1}{(1-z)^a}$$

$$_2F_1\!\left[\begin{matrix} a, 1-a \\ b \end{matrix}; \frac{1}{2}\right] = \frac{\Gamma(\tfrac{1}{2})\Gamma(\tfrac{1}{2}+a+b)}{\Gamma(\tfrac{1}{2}+a)\Gamma(\tfrac{1}{2}+b)}$$

$$_3F_2\!\left[\begin{matrix} -2n, b, c \\ 1-b-2n, 1-c-2n \end{matrix}; 1\right] = \frac{(1)_{2n}(b)_n(c)_n(b+c)_{2n}}{(1)_n(b)_{2n}(c)_{2n}(b+c)_n}$$

$$_3F_2\!\left[\begin{matrix} a, b, c \\ 1+a-b, 1+a-c \end{matrix}; 1\right] = \frac{(a-b)!\,(a-c)!\,(\tfrac{a}{2})!\,(\tfrac{a}{2}-b-c)!}{a!\,(\tfrac{a}{2}-b)!\,(\tfrac{a}{2}-c)!\,(a-b-c)!}$$

$$_3F_2\!\left[\begin{matrix} a, b, -n \\ 1+a-b, 1+a+n \end{matrix}; 1\right] = \frac{(1+a)_n(1+\tfrac{a}{2}-b)_n}{(1+\tfrac{a}{2})_n(1+a-b)_n}$$

$$_3F_2\!\left[\begin{matrix} a, b, c \\ \frac{1+a+b}{2}, 2c \end{matrix}; 1\right] = \frac{\Gamma(\tfrac{1}{2})\Gamma(c+\tfrac{1}{2})\Gamma(\tfrac{1}{2}+\tfrac{a}{2}+\tfrac{b}{2})\Gamma(\tfrac{1}{2}-\tfrac{a}{2}-\tfrac{b}{2}+c)}{\Gamma(\tfrac{1}{2}+\tfrac{a}{2})\Gamma(\tfrac{1}{2}+\tfrac{b}{2})\Gamma(\tfrac{1}{2}-\tfrac{a}{2}+c)\Gamma(\tfrac{1}{2}-\tfrac{b}{2}+c)}$$

$$_3F_2\!\left[\begin{matrix} a, 1-a, c \\ d, 1+2c-d \end{matrix}; 1\right] = \frac{\pi 2^{1-2c}(d-1)!\,(2c+d)!}{(\frac{a-d-1}{2})!\,(\frac{a+d}{2}-1)!\,(c-\frac{a+d}{2})!\,(\frac{d-a-1}{2})!}$$

3.6   Using the database

Let's review where we are. In this chapter we have seen how to take a sum and identify it, when possible, as a standard hypergeometric sum. We have also seen a list of many of the important hypergeometric sums that can be expressed in simple, closed form. We will now give a few examples of the whole process whereby one uses the hypergeometric database in order to try to "do" a given sum. The strengths and the limitations of the procedure should then be clearer.

Example 3.6.1.   For a nonnegative integer n, consider the sum

$$f(n) = \sum_k (-1)^k \binom{2n}{k}^2.$$

Can this sum be evaluated in some simple form?

The first step is to identify the sum $f(n)$ as a particular hypergeometric series. For that purpose we look at the term ratio

$$\frac{t_{k+1}}{t_k} = \frac{(-1)^{k+1}\binom{2n}{k+1}^2}{(-1)^k\binom{2n}{k}^2} = -\frac{(k-2n)^2}{(k+1)^2}.$$

Our series is thereby unmasked: it is

$$f(n) = {}_2F_1\left[\begin{matrix} -2n & -2n \\ & 1 \end{matrix}; -1\right].$$

Next we check the database to see if we have any information about ${}_2F_1$'s whose argument is -1, and, indeed, there is such an entry in the database, namely Kummer's identity. If we use it in the form that is given there for negative integer b, then it tells us that our unknown sum $f(n)$ is

$$f(n) = {}_2F_1\left[\begin{matrix} -2n & -2n \\ & 1 \end{matrix}; -1\right] = \frac{(2n-1)!2(-1)^n}{(n-1)!\,n!} = (-1)^n\binom{2n}{n},$$

which is a happy ending indeed. □

Example 3.6.2. For one more example of a lookup in the hypergeometric database, consider the sum

$$f(n) = \sum_k (-1)^k \binom{2n}{k}\binom{2k}{k}\binom{4n-2k}{2n-k}. \tag{3.6.1}$$

The first step is, as always, to find the term ratio and resolve it into linear factors. We find that

$$\frac{t_{k+1}}{t_k} = \frac{(-1)^{k+1}\binom{2n}{k+1}\binom{2k+2}{k+1}\binom{4n-2k-2}{2n-k-1}}{(-1)^k\binom{2n}{k}\binom{2k}{k}\binom{4n-2k}{2n-k}} = \frac{(k+\frac{1}{2})(k-2n)^2}{(k+1)^2(k-2n+\frac{1}{2})}.$$

The first term of our sum is not 1, but is instead $\binom{4n}{2n}$, hence we now know that our desired sum is

$$f(n) = \binom{4n}{2n}{}_3F_2\left[\begin{matrix} -2n & -2n & \frac{1}{2} \\ 1 & -2n+\frac{1}{2} \end{matrix}; 1\right]. \tag{3.6.2}$$

Since this is a ${}_3F_2$, we check the possibilities of Saalschütz's identity and Dixon's identity. It does not match the condition $d + e = a + b + c + 1$ of Saalschütz, so we try Dixon's next. If we put $a = -2n$, $b = -2n$, $c = \frac{1}{2}$, $d = 1$ and $e = -2n + \frac{1}{2}$, then we find that the conditions of Dixon's identity, namely that $d = a - b + 1$ and $e = a - c + 1$ are met. If we now use the right hand side of Dixon's identity, we find that if n is a nonnegative integer, then

$$f(n) = \binom{4n}{2n}\frac{(-n)!\,(-2n-\frac{1}{2})!\,(n-\frac{1}{2})!}{(-2n)!\,n!\,(-n-\frac{1}{2})!\,(-\frac{1}{2})!}.$$

This is a rather distressing development. We were expecting an answer in simple form, but the answer that we are looking at contains some factorials of negative numbers, and some of these negative numbers are negative *integers*, which are precisely the places where the factorial function is undefined.

Fortunately, what we have is a *ratio* of two factorials at negative integers; if we take an appropriate limit, the singularities will cancel, and a pleasant limiting ratio will ensue. We will now do this calculation, urging the reader to take note of the fact that this kind of situation happens fairly frequently when one uses the database. The answers are formally correct, but we need some further analysis to transform them into readily useable form.

What about the ratio $(-n)!/(-2n)!$? Imagine, for a moment, that n is near a positive integer, but is not equal to a positive integer. Then we use the reflection formula for the Γ-function

$$\Gamma(z)\Gamma(1-z) = \frac{\pi}{\sin \pi z}$$

once more, in the equivalent form

$$(-z)! = \frac{\pi}{(z-1)! \sin \pi z}.$$

When n is near, but not equal to, a positive integer we find that

$$\frac{(-n)!}{(-2n)!} = \frac{\pi}{(\sin n\pi)(n-1)!} \frac{(\sin 2n\pi)(2n-1)!}{\pi} = \frac{2(2n-1)! \cos n\pi}{(n-1)!}.$$

Thus as n approaches a positive integer, we have found that

$$\frac{(-n)!}{(-2n)!} \longrightarrow (-1)^n \frac{(2n)!}{n!}.$$

Our answer now has become

$$f(n) = \binom{4n}{2n} \frac{(-1)^n (2n)! \left(-2n-\frac{1}{2}\right)! \left(n-\frac{1}{2}\right)!}{n!^2 \left(-n-\frac{1}{2}\right)! \left(-\frac{1}{2}\right)!}.$$

A similar argument shows that

$$\frac{\left(-2n-\frac{1}{2}\right)!}{\left(-n-\frac{1}{2}\right)!} = \frac{(-1)^n \left(n-\frac{1}{2}\right)!}{\left(2n-\frac{1}{2}\right)!},$$

which means that now

$$f(n) = \binom{4n}{2n}\binom{2n}{n} \frac{\left(n-\frac{1}{2}\right)!^2}{\left(2n-\frac{1}{2}\right)! \left(-\frac{1}{2}\right)!}.$$

But for every positive integer m,

$$(m - \frac{1}{2})! = (m - \frac{1}{2})(m - \frac{3}{2}) \cdots (\frac{1}{2})(-\frac{1}{2})!$$

$$= \frac{(2m-1)(2m-3)\cdots 1}{2^m}(-\frac{1}{2})!$$

$$= \frac{(2m)!}{4^m m!}(-\frac{1}{2})!.$$

So we can simplify our answer all the way down to $f(n) = \binom{2n}{n}^2$. The fruit of our labor is that we have found the identity

$$\sum_k (-1)^k \binom{2n}{k}\binom{2k}{k}\binom{4n-2k}{2n-k} = \binom{2n}{n}^2; \qquad (3.6.3)$$

we realize that it is a special case of Dixon's identity, and we further realize that the "lookup" in the database was not quite a routine matter! □

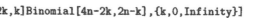

Since that was a very tedious lookup operation, might one of our computer packages have been able to help? Indeed, in Mathematica, the SymbolicSum package can handle the sum (3.6.1) easily. If we read in the package with

```
<<Algebra'SymbolicSum'
```

and then call for

```
Sum[(-1)^kBinomial[2n,k] Binomial[2k,k] Binomial[4n-2k,2n-k],{k,0,2n}]
```

we obtain the reply

$$\frac{(-4)^n (-1)^n (2n)! (4n)! \, \mathrm{Gamma}(\frac{1}{2} - 2n)}{4^n 16^n \sqrt{\pi}\, n!^4}.$$

This is easily seen to be the same as the evaluation (3.6.3).

The *Hyp* package includes a large database, considerably larger than the list that we have given above. So let's see how it does with the same summation problem.

Hence enter

```
Sum[(-1)^kBinomial[2n,k]Binomial[2k,k]Binomial[4n-2k,2n-k],{k,0,Infinity}]
```

and then request % //.SumF. The resulting output is exactly as in (3.6.2). So we have successfully identified the sum as a hypergeometric series.

The next question is, does *Hyp* know how to evaluate this sum in simple form? To ask Hyp to look up your sum in its sum list we use the command SListe. More precisely, we ask it to apply the rule SListe to the previous output by typing

```
% /.SListe
```

It replies by giving the numbers of the formulas in its database that might be of assistance in evaluating our sum. In this case its reply is to tell us that one of its four items S3202, S3231, S3232, S3233 might be of use. Let's now find out if item S3202 really will help, which we do by entering `mysum/.S3202`. Its answer to that is indeed the evaluation of our sum, in the form

$$\frac{(\frac{1}{2})_n(\frac{1}{2}-2n)_{2n}(-2n)_n(1+2n)_{2n}}{(\frac{1}{2})_{2n}(1)_n(\frac{1}{2}-2n)_n(-2n)_{2n}},$$

which is a rather nontransparent form of $\binom{2n}{n}^2$, but we mustn't begrudge the small amount of tidying up that we have to do, considering the lengthy limiting arguments that we have avoided.

3.7   Is there really a hypergeometric database?

Ask any computer scientist what a database is, and you will be told something like this. A database $\mathcal{D}$ is a triple consisting of

1. a collection of information (data) and

2. a collection of *queries* (questions) that may be addressed to the database $\mathcal{D}$ by the user and

3. a collection of algorithms by which the system responds to queries and searches the data in order to find the answers.

There is no hypergeometric database. It's a myth.

Is there a collection of information? The data might be, for instance, a list of all known hypergeometric identities. But there isn't any such list. If you propose one, somebody will produce a known identity that isn't on your list. But suppose that problem didn't exist. Let's compromise a bit, and settle for a very large collection of many of the most important hypergeometric identities.

Fine. Now what are the queries that we would like to address to the database? That's a lot easier. Suppose there is just one query: "Is the following sum expressible in simple terms, and if so, in what simple terms?"

All right. We're trying to construct a collection of identities that will be equipped to discover if somebody's sum can or cannot be expressed in a much simpler form.

We're two-thirds of the way there. We have a (slightly mythical) collection of data, and a single rather precise query. What we are missing is the algorithm.

If some user asks the system whether or not a certain sum can be expressed in some simple form, exactly what steps shall the system take in order to answer the question?

Certainly a minimal step would be to examine the list of known identities and see if the sum in question lives there. The hypergeometric notation that we have described in this chapter will be a big help in doing that search. If the sum, exactly as described by the user, does reside in the data, then we're finished. The system will simply print out the simple evaluation, and the user will go away with a smile.

But suppose the sum does not appear among the data in the system? Well, you might say that the system should apologize, and declare that it can't help. That would be honest, but not very useful. In fact, the act of checking whether the given sum lives among the data is only the first step that any competent human analyst would take. If the sum could not be located, the next step that the analyst would probably take would be to try out some hypergeometric *transformation rules*.

A transformation rule is a relation between two hypergeometric functions whose parameter sets are different, which shows, nonetheless, that if you know one of them then you know both of them. Here is one of the many, many known hypergeometric transformation rules:

$$_2F_1\left[\begin{matrix} a, b \\ c \end{matrix}; z\right] = (1-z)^{c-a-b}\,_2F_1\left[\begin{matrix} c-a, c-b \\ c \end{matrix}; z\right].$$

That is a rule, an identity really, that relates two $_2F_1$'s with different parameter lists. One could easily make lists of dozens of such rules, and indeed the package *Hyp* has dozens of them built in.

So your database should first look to see if your sum lives in the data, and if not it should next try to transform your sum into another one that does live in the data. If that succeeds, great. If it fails, well maybe there's a sequence of two transformations that will do it. Or maybe three — you see the problem. Besides sequences of transformations, one can also use various substitutions for the parameters, and it may be hard to recognize that a certain identity is a specialization of a database entry.

There is no algorithm that will discover whether your sum is or is not transformable into an identity that lives in the database.

Beyond all of these attempts at computer algorithms there lie human mathematicians. Many of them are awesomely bright, and will find immensely clever ways to evaluate your unknown sum, ways that could not in a million years be built into a computerized database.

So the database of hypergeometric identities is a myth. It is very nice to have a big list to work with. But that is by no means the whole story.

The look-it-up-in-a-database process is, like any other, an *algorithm* for doing hypergeometric sums, and it should be assessed the same way as other algorithms. How effective is it? Precisely when can we expect a pretty answer? How fast is it? What is the *complete* algorithm, including the simplifications at the end, and how costly are they? What are the alternatives?

In the sequel we will present other algorithms, ones that don't involve any lookup in or manipulation of a database, for doing hypergeometric sums in simple form. Those algorithms can be rather easily programmed for a computer, they work under conditions that are wider than those of the database lookup, and the conditions under which they work can be clearly stated. Further, under the stated conditions these algorithms are exhaustive. That is, if they produce nothing, then that is a proof that nothing exists, rather than only a confession of possible inadequacy.

Obviously we cannot claim that the computerized methods are the best for every situation. Sometimes the certificates that they produce are longer and less user-friendly than those that humans might find, for example. But the emergence of these methods has put an important family of tools in the hands of discrete mathematicians, and many results that are accessible in no other way have been found and proved by computer methods.

3.8 Exercises

1. Put each of the following sums into standard hypergeometric notation. First do it by hand. Then do it with your choice of computer software.

   (a) $\sum_k \binom{n}{k} \binom{n+a}{k} \binom{n-a}{n-k}$

   (b) $\sum_k \binom{n}{k}^p$

   (c) $\sum_k (-1)^k \binom{k}{n-k}$

2. For selected entries (of your choice) in the hypergeometric database list in this chapter, express the summand using only

   (a) rising factorials

   (b) the gamma function

   (c) factorials

   (d) binomial coefficients (when is this possible?)

3. Each of the following sums can be evaluated in a simple form. In each case first write the sum in standard hypergeometric notation. Then consult the list in this chapter to find a database member that has the given sum as

a special case. Then use the right hand side of the database sum, suitably specialized, to find the simple form of the given sum. Then check your answer numerically for a few small values of the free parameters.

(a) $\sum_{k=0}^{n} \binom{n}{k} / \binom{2n-1}{k}$

(b) $\sum_{k} \binom{n}{k}^{2} \binom{3n+k}{2n}$

(c) $\sum_{k}(-1)^{k} \binom{n}{k} \frac{(k+3a)!}{(k+a)!}$

(d) $\sum_{k}(-1)^{k} \binom{k+a}{a} \binom{n+k}{2k+2a} 4^{k}$

(e) $\sum_{k} \binom{2n+2}{2k+1} \binom{x+k}{2n+1}$

(f) $\sum_{k} \binom{2n+1}{2p+2k+1} \binom{p+k}{k}$

(g) $\sum_{k}(-1)^{k} \binom{2n}{k} \binom{2x}{x-n+k} \binom{2z}{z-n+k}$

Part II

The Five Basic Algorithms

Chapter 4

Sister Celine's Method

4.1 Introduction

The subject of computerized proofs of identities begins with the Ph.D. thesis of Sister Mary Celine Fasenmyer at the University of Michigan in 1945. There she developed a method for finding recurrence relations for hypergeometric polynomials directly from the series expansions of the polynomials. An exposition of her method is in Chapter 14 of Rainville [Rain60]. In his words,

> Years ago it seemed customary upon entering the study of a new set of polynomials to seek recurrence relations ... by essentially a hit-or-miss process. Manipulative skill was used and, if there was enough of it, some relations emerged; others might easily have been lurking around a corner without being discovered ... The interesting problem of the pure recurrence relation for hypergeometric polynomials received probably its first systematic attack at the hands of Sister Mary Celine Fasenmyer ...

The method is quite effective and easily computerized, though it is usually slow in comparison to the methods of Chapter 6. Her algorithm is also important because it has yielded general existence theorems for the recurrence relations satisfied by hypergeometric sums.

We begin by illustrating her method on a simple sum.

Example 4.1.1. Let

$$f(n) = \sum_k k \binom{n}{k} \qquad (n = 0, 1, 2, \ldots),$$

and let's look for the recurrence that $f(n)$ satisfies. To do this we first look for the recurrence that the *summand*

$$F(n, k) = k\binom{n}{k}$$

satisfies. It is a function of two variables (n, k), so we try to find a recurrence of the form

$$a(n)F(n, k) + b(n)F(n + 1, k) + c(n)F(n, k + 1) + d(n)F(n + 1, k + 1) = 0, \quad (4.1.1)$$

in which the coefficients a, b, c, d *depend on n only, and not on k* (for reasons that will become clear presently).

To find the coefficients, if they exist, we divide (4.1.1) through by $F(n, k)$ getting

$$a + b\frac{F(n + 1, k)}{F(n, k)} + c\frac{F(n, k + 1)}{F(n, k)} + d\frac{F(n + 1, k + 1)}{F(n, k)} = 0. \qquad (4.1.2)$$

Now each of the ratios of F's is a certain rational function. If we carry out the indicated divisions, our assumed recurrence becomes

$$a + b\frac{n + 1}{n + 1 - k} + c\frac{n - k}{k} + d\frac{n + 1}{k} = 0,$$

in which the factorials have all disappeared, and we see only rational functions of n and k.

The next step, following Sister Celine, is to put the whole thing over a common denominator. That denominator is $k(n + 1 - k)$ and the numerator is, after collecting by powers of k,

$$(d + (c + 2d)n + (c + d)n^2) + (a + b - c - d + (a + b - 2c - d)n)k + (c - a)k^2.$$

Our assumed recurrence will be true if this numerator identically vanishes, for all n and k, with the coefficients a, b, c, d being permitted to depend only on n. Thus the coefficient of each power of k must vanish. This gives us a system of three equations in four unknowns, namely

$$\begin{bmatrix} 0 & 0 & n(n + 1) & (n + 1)^2 \\ n + 1 & n + 1 & -2n - 1 & -(n + 1) \\ -1 & 0 & 1 & 0 \end{bmatrix} \begin{bmatrix} a \\ b \\ c \\ d \end{bmatrix} = \begin{bmatrix} 0 \\ 0 \\ 0 \end{bmatrix},$$

to solve for a, b, c, d.

Success in finding a nontrivial solution is now guaranteed simply because there are more unknowns than there are equations. If we actually solve these equations we find that

$$[a, b, c, d] = d\left[-1 - 1/n \quad 0 \quad -1 - 1/n \quad 1\right].$$

We now substitute these values into the assumed form of the recurrence relation in (4.1.1), and we have the desired "k-free" recurrence for the summand $F(n, k)$, namely

$$-(1 + \frac{1}{n})F(n, k) - (1 + \frac{1}{n})F(n, k + 1) + F(n + 1, k + 1) = 0. \qquad (4.1.3)$$

From the recurrence for the *summand* $F(n, k)$ to the recurrence for the *sum* $f(n)$ is a very short step: just sum (4.1.3) over *all integers* k, noticing appreciatively that the coefficients in the recurrence are free of k's, so the summation over k can operate directly on the F in each term. We get instantly the recurrence

$$-(1 + \frac{1}{n})f(n) - (1 + \frac{1}{n})f(n) + f(n + 1) = 0,$$

i.e., the recurrence

$$f(n + 1) = 2\frac{n + 1}{n}f(n) \qquad (n = 1, 2, \ldots; f(1) = 1).$$

We can now easily find $f(n)$, the desired sum, since

$$f(n + 1) = 2\frac{n + 1}{n}f(n) = 2^2\frac{n + 1}{n}\frac{n}{n - 1}f(n - 1) = \cdots = 2^n(n + 1)f(1),$$

so

$$f(n) = n2^{n-1}$$

for all $n \geq 0$. □

Yes, this was a very simple example, but it illustrated many of the points of interest. The method works because one can prove (and we will!) that the number of unknowns can always be made larger than the number of equations, so a nontrivial solution must exist. The fact that the coefficients in the assumed recurrence are free of k is vital to the step in which we sum the F recurrence over all integers k, as we did in (4.1.3).

4.2   Sister Mary Celine Fasenmyer

Sister Celine was born in Crown, in central Pennsylvania, October 4, 1906. Her parents were George and Cecilia Fasenmyer, though her mother, Cecilia, died when Mary was one year old. Her father worked his own oil lease in the area. He remarried three years later a woman, Josephine, who was twenty-five years his junior.

Mary's early education was at the St. Joseph's Academy in Titusville, Pennsylvania, from which she was graduated in 1923, having been always "good in math." She then taught for ten years, and in 1933 received her AB degree from Mercyhurst College. She was sent to Pittsburgh by her order, to teach in the St. Justin School and to go to the University of Pittsburgh for her MA degree, which she received in 1937. Her major was mathematics, and her minor was in physics. The community told her to go to the University of Michigan for her doctorate, which she did from the fall of 1942 until June of 1946, when she received her degree. Her thesis was written under the direction of Earl Rainville, whom she remembers as having been quite accessible and helpful, as well as working in a subject area that she liked. In her thesis she showed how one can find recurrence relations that are satisfied by sums of hypergeometric terms, in a purely mechanical ("algorithmic") way. She used the method in her thesis [Fase45] to find pure recurrence relations that are satisfied by various hypergeometric polynomial sequences. In two later papers she developed the method further, and explained its workings to a broad audience in her paper [Fase49]. For an exposition of some of her thesis results see [Fase47]. Her work was described by Rainville in Chapters 14 and 18 of his book [Rain60]. Her method is the intellectual progenitor of the computerized methods that we use today to find and prove hypergeometric identities, thanks to the recognition that it can be adapted to prove such identities via Zeilberger's paradigm (see page 23).

4.3   Sister Celine's general algorithm

Now let's discuss her algorithm in general. We are given a sum $f(n) = \sum_k F(n, k)$, where F is doubly hypergeometric. That is, both

$$F(n+1, k)/F(n, k) \text{ and } F(n, k+1)/F(n, k)$$

are rational functions of n and k. We want to find a recurrence formula for the sum $f(n)$, so for a first step, we will find a recurrence for the *summand* $F(n, k)$, of the form

$$\sum_{i=0}^{I} \sum_{j=0}^{J} a_{i,j}(n) F(n-j, k-i) = 0. \tag{4.3.1}$$

The complete sequence of steps is the following.

1. Fix trial values of I and J, say $I = J = 1$.

2. Assume the recurrence formula in the form of (4.3.1), with the coefficients $a_{ij}(n)$ to be determined, if possible.

3. Divide each term of (4.3.1) by $F(n,k)$, and reduce each ratio $F(n-j,k-i)/F(n,k)$ by simplifying the ratios of the factorials that it contains, so that only rational functions of n and k remain.

4. Place the entire expression over a single common denominator. Then collect the numerator as a polynomial in k.

5. Solve the system of linear equations that results from equating to zero the coefficients of each power of k in the numerator polynomial, for the unknown coefficients $a_{i,j}$. If the system has no solution, try the whole thing again with larger values of I and/or J. That is, look for a bigger recurrence. $\square$

We will prove below that under suitable hypotheses Sister Celine's algorithm is guaranteed to succeed if I, J are large enough, and the "large enough" can be estimated in advance. But first let's look at implementations of her algorithm in Maple and Mathematica.

The Maple program for her algorithm is contained in the package EKHAD that is included with this book (see Appendix A). To use it just call celine(f,ii,jj);, where f is your summand, and ii, jj are the sizes of the recurrence that you are looking for.

In the above example, we would call celine((n,k) -> k*n!/(k!*(n-k)!),1,1); and we would obtain the following output:

```
The full recurrence is
   b[3] n F(n-1,k)-(n-1) b[3] F(n,k)+b[3] F(n-1,k-1) n== 0
```

In this output b[3] is an arbitrary constant, which can be ignored. The recurrence is identical with the one we had previously found by hand, in (4.1.3), as can be seen by replacing n and k by $n-1$ and $k-1$ in (4.1.3) and comparing it with the output above.

Example 4.3.1. Suppose we were to use the program with the input $f(n,k) = \binom{n}{k}$. Then what recurrence would it find? If you guessed the Pascal triangle recurrence $\binom{n}{k} = \binom{n-1}{k} + \binom{n-1}{k-1}$, then you would be right. In that same spirit, let's look for a recurrence that $f(n,k) = \binom{n}{k}^2$ satisfies.

So we now call `celine((n, k) → (n!/(k! * (n − k)!))², 2, 2);`. The program runs for a while, and then announces that

```
The full recurrence is
  (n - 1) b[8] F(n - 2, k - 2) + b[8] (2 - 2 n) F(n - 2, k - 1)
    + (- 2 n + 1) b[8] F(n - 1, k - 1) + (n - 1) b[8] F(n - 2, k)
    + (- 2 n + 1) b[8] F(n - 1, k) + b[8] n F(n, k) == 0
```

Translated into conventional mathematical notation, this recurrence reads as

$$n\binom{n}{k}^2 - (2n-1)\left\{ \binom{n-1}{k}^2 + \binom{n-1}{k-1}^2 \right\}$$
$$+ (n-1)\left\{ \binom{n-2}{k}^2 - 2\binom{n-2}{k-1}^2 + \binom{n-2}{k-2}^2 \right\} = 0, \qquad (4.3.2)$$

which is the "Pascal triangle" identity for the squares of the binomial coefficients.

Let's use the recurrence for the squares to find a recurrence for the *sum* of the squares. So let $f(n) = \sum_k \binom{n}{k}^2$. What is the recurrence for $f(n)$? To find it, just sum (4.3.2) over all integers k, obtaining

$$nf(n) - (2n-1)\{f(n-1) + f(n-1)\} + (n-1)\{f(n-2) - 2f(n-2) + f(n-2)\} = 0$$

which boils down to just

$$f(n) = \frac{2(2n-1)}{n}f(n-1) = \frac{2^2(2n-1)(2n-3)}{n(n-1)}f(n-2) = \cdots = \frac{(2n)!}{n!^2}.$$

We therefore have another illustration of how the computer can discover the evaluation in closed form of a hypergeometric sum.   □

Cubes and higher powers of the binomial coefficients also satisfy recurrences of this kind, but finding them with this program would require either an immense computer or immense patience. The sums of the cubes, for instance, of the binomial coefficients also satisfy a recurrence, which the method would discover. That recurrence is of second order, however, and its solution *provably* cannot be expressed as a linear combination of a constant number of hypergeometric terms (see Chapter 8). Hence, in the case of the sum of the cubes, what we get is a computer generated proof of the impossibility of finding a pleasant evaluation, with a reasonable definition of "pleasant." The theory of this remarkable chain of recent developments will be fully explained later (see Section 8.6 and Theorem 8.8.1).

Next let's do the same job in Mathematica. The first thing to do is to read in the package `DiscreteMath'RSolve'` in order to enable the `FactorialSimplify` instruction. Next we define a Module that finds a recurrence relation satisfied by a given function f, the recurrence being of orders ii, jj.

```
<<DiscreteMath'RSolve'
findrecur[f_ ,ii_ ,jj_ ]:=
  Module[{yy,zz,ll,tt,uu,r,s,i,j},
    yy=Sum[Sum[a[i,j] FactorialSimplify[f[n-j,k-i]/f[n,k]],
      {i,0,ii}],{j,0,jj}];
    zz=Collect[Numerator[Together[yy]],k];
    ll=CoefficientList[zz,k];
    tt=Flatten[Table[a[i,j],{i,0,ii},{j,0,jj}]];
    uu=Flatten[Simplify[Solve[ll==0,tt]]];
    For[r=0,r<=ii,r++,
      For[s=0,s<=jj,s++,
        a[r,s]=Replace[a[r,s],uu]]];
    Sum[Sum[a[i,j] F[n-j,k-i],{i,0,ii}],{j,0,jj}]==0]
```

After that there's nothing to do but define the function

```
    f[n_ ,k_ ]:=k n!/(k!(n-k)!)
```

and call the Module

```
    findrecur[f,1,1].
```

The resulting output is

```
    a[1,1] F[-1+n,-1+k] + a[1,1] F[-1+n,k] + (-1+1/n) a[1,1] F[n,k]==0,
```

as before.

The reader is cautioned that this program is slower than its Maple counterpart in its execution, and it should not be tried on recurrence relations of larger span.

Next let's try an example in the spirit of Sister Celine's original use of the recurrence method.

Example 4.3.2. We'll look for a recurrence that is satisfied by the classical Laguerre polynomials

$$L_n(x) = \sum_{k=0}^{n} (-1)^k \binom{n}{k} \frac{x^k}{k!} \qquad (n = 0, 1, 2, \ldots).$$

The first step is to find a two-variable recurrence that is satisfied by the summand itself, in this case by

$$F(n,k) = (-1)^k \binom{n}{k} \frac{x^k}{k!} \qquad (n, k \geq 0). \qquad (4.3.3)$$

The "Fundamental Theorem," Theorem 4.4.1 below, guarantees that this F satisfies a recurrence. Before we go to the computer to find the recurrence, let's try to estimate its order in advance. To do this, identify the specific F in (4.3.3) with the general form in (4.4.1) below by taking $uu := 1$, $vv := 3$, $x := -x$, $(a_1, b_1, c_1) = (1, 0, 0)$ and for the three (u, v, w) vectors,

$$(0, 1, 0), \ (1, -1, 0), \ (0, 1, 0).$$

Then for the quantitative estimates provided by the theorem, namely the (I^*, J^*) of (4.4.3), we find $J^* = 3$, $I^* = 4$. Hence there is surely a recurrence for $F(n, k)$ of the form

$$\sum_{i=0}^{4} \sum_{j=0}^{3} a_{i,j}(n) F(n - j, k - i) = 0,$$

in which the $a_{i,j}$'s are polynomials in n, and are not all zero.

To find such a recurrence we use the Maple program `celine` above. To search for a recurrence of orders 2, the program will be called with

```
celine((n,k) -> (-1)^k*n!*x^k/(k!^2*(n-k)!),2,2);
```

The program runs briefly, and returns the following output:

```
The full recurrence is
(-b[3]*x+b[3]*x*n)*F(n-2,k-1)+(n**2*b[4]-n*b[4]+b[3]+2*b[3]*n**2
   -3*b[3]*n)*F(n-2,k)+(-2*n**2*b[4]+4*b[3]*n-b[3]+n*b[4]
   -4*b[3]*n**2)*F(n-1,k)+(n**2*b[4]+2*b[3]*n**2-b[3]*n)*F(n,k)
   +b[4]*F(n-1,k-1)*x*n+b[3]*x**2*F(n-1,k-2)+b[3]*F(n,k-1)*x*n==0
```

in which $b[3]$, $b[4]$ are arbitrary constants.

But the recurrence for $F(n, k)$ wasn't the object of the exercise. What we wanted was a recurrence for the Laguerre polynomial

$$L_n(x) = \sum_k F(n, k),$$

in which the sum on k is over *all* integers, i.e., extends from $-\infty$ to $+\infty$. This point is extremely important. The summand $F(n, k)$ of (4.3.3) will vanish automatically if $k < 0$ or if $k > n$, i.e., it has *compact support*. Hence even if we sum over all integers, the sums will contain only finitely many nonvanishing terms.

Therefore, we can sum the output recurrence over *all integers* k. Further, since the constants $b[3], b[4]$ are arbitrary, let's take $b[3] = 0$ and $b[4] = 1$. This yields

$$(n^2 - n)L_{n-2}(x) + (-2n^2 + n)L_{n-1}(x) + n^2 L_n(x) + nx L_{n-1}(x) = 0$$

or finally

$$nL_n(x) + (x + 1 - 2n)L_{n-1}(x) + (n - 1)L_{n-2}(x) = 0,$$

which is a well known three term recurrence relation for the Laguerre polynomials.
□

It should be noted that while the Laguerre polynomials make a pleasant example, the main theorem, Theorem 4.4.1, which is stated and proved in the next section, assures us that *every* sequence of polynomials $\sum_k F(n, k)x^k$, where F is a proper hypergeometric term, satisfies a recurrence relation, and it even gives us a bound on the order of the recurrence.

Example 4.3.3. In this example we will see that with Sister Celine's original algorithm we can easily find the right hand sides of some fairly formidable identities. Consider the evaluation of the sum

$$f(n) = \sum_k \binom{n}{k}\binom{2k}{k}(-2)^{n-k}. \tag{4.3.4}$$

Without thinking, just enter the summand $F(n, k)$ into the input line of program celine as

```
celine((n,k) -> n!*(2*k)!*(-2)^(n-k)/(k!^3*(n-k)!),2,2);
```

The output is as follows.

```
The full recurrence is
(-8*b[3]*n+8*b[3])*F(n-2,k-2)+(2*b[3]-4*b[3]*n)*F(n-1,k-2)
  +(-8*b[0]*n+8*b[0]+4*b[3]*n-4*b[3])*F(n-2,k-1)
  +(4*b[3]*n-2*b[3]-4*b[0]*n+2*b[0])*F(n-1,k-1)
  +(4*b[0]*n-4*b[0])*F(n-2,k)+(4*b[0]*n-2*b[0])*F(n-1,k)
  +b[0]*F(n,k)*n+b[3]*F(n,k-1)*n ==0
```

Since b[0] and b[3] are arbitrary constants here, we might as well choose, say, b[0]=1 and b[3]=0, in which case the above recurrence simplifies to

$$-8(n-1)F(n-2, k-1) - 2(2n-1)F(n-1, k-1)$$
$$+ 4(n-1)F(n-2, k) + 2(2n-1)F(n-1, k) + nF(n, k) = 0.$$

If we now sum over all integers k, we find that the sum $f(n)$, of (4.3.4), satisfies the beautifully simple recurrence

$$nf(n) - 4(n-1)f(n-2) = 0.$$

Since, from (4.3.4), $f(0) = 1$ and $f(1) = 0$, it follows immediately that

$$f(n) = \begin{cases} 0, & \text{if } n \text{ is odd;} \\ \binom{n}{n/2}, & \text{if } n \text{ is even.} \end{cases}$$

The above is called the Reed–Dawson identity, and Sister Celine's algorithm *derived and proved* it effortlessly. $\square$

4.4   The Fundamental Theorem

The "Fundamental Theorem" states that every proper hypergeometric term $F(n, k)$ satisfies a recurrence relation of the kind we have found in the previous chapters, and it validates the procedure that we have used to find these recurrences in the sense that it guarantees that Sister Celine's method will work if the span of the assumed recurrence is large enough. The theorem also finds explicit precomputable upper bounds on the span.

Definition. A function $F(n, k)$ is said to be a *proper hypergeometric term* if it can be written in the form

$$F(n, k) = P(n, k) \frac{\prod_{i=1}^{uu} (a_i n + b_i k + c_i)!}{\prod_{i=1}^{vv} (u_i n + v_i k + w_i)!} x^k, \tag{4.4.1}$$

in which x is an indeterminate over, say, the complex numbers, and

1. P is a polynomial,

2. the a's, b's, u's, v's are specific integers, that is to say, they do not contain any additional parameters, and

3. the quantities uu and vv are finite, nonnegative, specific integers. $\square$

An F of the form (4.4.1) is *well defined* at a point (n, k) if none of the numbers $\{a_i n + b_i k + c_i\}_{i=1}^{uu}$ is a negative integer. We will say that $F(n, k) = 0$ if F is well defined at (n, k) and at least one of the numbers $\{u_i n + v_i k + w_i\}_{i=1}^{vv}$ is a negative integer, or $P(n, k) = 0$.

Some examples of proper hypergeometric terms are as follows.

The term $\binom{n}{k} 2^k$ is proper hypergeometric because it can be written

$$F(n, k) = \binom{n}{k} 2^k = \frac{n!}{k! (n-k)!} 2^k,$$

which is exactly of the required form. Also, $\binom{2n}{k} 2^k$ is proper hypergeometric, but $\binom{an}{n}$ is not, if a is an unspecified parameter.

Consider, however, $F(n, k) = 1/(n + 3k + 1)$. This is not in proper hypergeometric form. It doesn't contain any of the factorials, and it isn't a polynomial.

However, the definition says "... if it can be written in the form ..." This $F(n, k)$ can be written in proper hypergeometric form, even though it was not given to us in that form! All we have to do is to write

$$\frac{1}{n + 3k + 1} = \frac{(n + 3k)!}{(n + 3k + 1)!}.$$

On the other hand, if we take $F(n, k) = 1/(n^2 + k^2 + 1)$, then no amount of rewriting will produce the form in (4.4.1), so this F is not proper hypergeometric.

Now we can state the main theorem.

Theorem 4.4.1 *Let $F(n, k)$ be a proper hypergeometric term. Then F satisfies a k-free recurrence relation. That is to say, there exist positive integers I, J, and polynomials $a_{i,j}(n)$ for $i = 0, \ldots, I; j = 0, \ldots, J$, not all zero, such that the recurrence*

$$\sum_{i=0}^{I} \sum_{j=0}^{J} a_{i,j}(n) F(n - j, k - i) = 0 \qquad (4.4.2)$$

holds at every point (n, k) at which $F(n, k) \neq 0$ and all of the values of F that occur in (4.4.2) are well defined. Furthermore, there is such a recurrence with $(I, J) = (I^*, J^*)$ where

$$J^* = \sum_s |b_s| + \sum_s |v_s|; \quad I^* = 1 + \deg(P) + J^* (\{\sum_s |a_s| + \sum_s |u_s|\} - 1). \quad (4.4.3)$$

Note that the recurrence (4.4.2) is *k-free* since the coefficients $a_{i,j}(n)$ depend only on n, not on k.

Next we're going to prove the theorem. In order to do that it will be important to do a few simple exercises that relate to the behavior of *translates of a proper hypergeometric term.*

Suppose $f(n) = (2n + 3)!$. What is $f(n - 2)/f(n)$? It is

$$\frac{f(n - 2)}{f(n)} = \frac{1}{4n(1 + n)(1 + 2n)(3 + 2n)},$$

i.e., it is the reciprocal of a certain polynomial in n.

On the other hand, if $f(n) = (3 - 2n)!$, then

$$\frac{f(n - 2)}{f(n)} = 4(n - 3)(n - 2)(2n - 7)(2n - 5)$$

is a polynomial in n.

The conclusion here is that if $f(n) = (an + b)!$, then the ratio $f(n - j)/f(n)$ is, for $j \geq 0$, a polynomial in n, if $a \leq 0$, and the reciprocal of a polynomial in n, if $a > 0$.

For a two-variable case, consider $f(n, k) = (2n - 3k + 1)!$. Then

$$\frac{f(n-9, k-5)}{f(n, k)} = \frac{1}{(-1 - 3k + 2n)(-3k + 2n)(1 - 3k + 2n)}$$

is the reciprocal of a polynomial in n and k, whereas

$$\frac{f(n-6, k-5)}{f(n, k)} = (2 - 3k + 2n)(3 - 3k + 2n)(4 - 3k + 2n)$$

is a polynomial in n and k.

The general rule in the two-variable case is that if $F(n, k) = (an + bk + c)!$ then for $i, j \geq 0$ we have $F(n - j, k - i)/F(n, k)$ equal to

$$\begin{cases} \{(an + bk + c) \cdots (an + bk + c - aj - bi + 1)\}^{-1}, & \text{if } aj + bi \geq 0; \\ (an + bk + c + |aj + bi|) \cdots (an + bk + c + 1), & \text{if } aj + bi < 0. \end{cases} \quad (4.4.4)$$

Once again the result is either a polynomial in n and k, or is the reciprocal of such a polynomial, depending on the sign of $aj + bi$.

Let's introduce, just for the purposes of this proof, the following notation for the rising factorial (rf) and falling factorial (ff) polynomials, for nonnegative integer values of x (the empty product is $=1$):

$$\text{rf}(x, y) = \prod_{j=1}^{x} (y + j),$$

$$\text{ff}(x, y) = \prod_{j=0}^{x-1} (y - j).$$

In terms of these polynomials, we can rewrite (4.4.4) as

$$\frac{F(n - j, k - i)}{F(n, k)} = \begin{cases} 1/\text{ff}(aj + bi, an + bk + c), & \text{if } aj + bi \geq 0; \\ \text{rf}(|aj + bi|, an + bk + c), & \text{if } aj + bi < 0. \end{cases} \quad (4.4.5)$$

Now consider a function $F(n, k)$ which is not just a single factor $(an + bk + c)!$, but is a product of several such factors divided by another product of several such factors, as in (4.4.1) above,

$$F(n, k) = P(n, k) \frac{\prod_{s=1}^{uu} (a_s n + b_s k + c_s)!}{\prod_{s=1}^{vv} (u_s n + v_s k + w_s)!} x^k.$$

What can we say about the form of the ratio $\rho = F(n - j, k - i)/F(n, k)$ now? Well, each of the factorial factors in the numerator of F contributes a polynomial in n and k either to the numerator of ρ or to the denominator of ρ, as in (4.4.5). Hence, the ratio ρ will be a rational function of n and k, say $\nu(n,k)/\delta(n,k)$.

More precisely, the numerator $\nu(n,k)$ of the ratio ρ will be, according to (4.4.5),

$$P(n-j,k-i)\prod_{\substack{s=1 \\ a_s j+b_s i<0}}^{uu} \mathrm{rf}(|a_s j+b_s i|, a_s n+b_s k+c_s)\prod_{\substack{s=1 \\ u_s j+v_s i\geq0}}^{vv} \mathrm{ff}(u_s j+v_s i, u_s n+v_s k+w_s),$$

$$(4.4.6)$$

and its denominator $\delta(n,k)$ will be

$$P(n,k)x^i \prod_{\substack{s=1 \\ a_s j+b_s i\geq0}}^{uu} \mathrm{ff}(a_s j+b_s i, a_s n+b_s k+c_s)\prod_{\substack{s=1 \\ u_s j+v_s i<0}}^{vv} \mathrm{rf}(|u_s j+v_s i|, u_s n+v_s k+w_s).$$

$$(4.4.7)$$

OK, now, to prove the theorem, let's assume the recurrence in the form (4.4.2) and try to solve for the coefficients $a_{i,j}(n)$. If we do that, then after dividing by $F(n,k)$, the left side of the assumed recurrence will be

$$\sum_{0\leq i\leq I; 0\leq j\leq J} a_{i,j}(n)\frac{\nu_{i,j}}{\delta_{i,j}}, \qquad (4.4.8)$$

where each $\nu_{i,j}$ looks like (4.4.6) and each $\delta_{i,j}$ looks like (4.4.7).

The next step is to collect all of the terms in the sum (4.4.8) over a single least common denominator. But since we have such explicit formulas for the numerator and denominator polynomials of each term, we can write out explicitly what that common denominator will be.

The first thing to notice is that, by (4.4.7), each and every denominator in (4.4.8) contains the same factor $P(n,k)$, so $P(n,k)$ will be in the least common denominator that we are constructing.

(Notice that if we had permitted a denominator polynomial $Q(n,k)$ to appear in the definition (4.4.1) of a proper hypergeometric term, then the outlook would have been much bleaker. Indeed, in that case, the (i,j) term in (4.4.8) would have contained a factor $Q(n-j,k-i)$ also, and we would need to deal with the least common multiple of all such factors when constructing the common denominator of all terms. That multiple would have been of an unacceptably high degree in k, and would have blocked the argument that follows from reaching a successful conclusion.)

We introduce the symbol $x^+ = \max(x,0)$, where x is a real number. Then for all real numbers a,b we have

$$\max\left\{|aj+bi| : aj+bi\leq0; 0\leq i\leq I; 0\leq j\leq J\right\} = (-a)^+ J + (-b)^+ I,$$

and
$$\max\left\{aj + bi : aj + bi \geq 0; 0 \leq i \leq I; 0 \leq j \leq J\right\} = a^+ J + b^+ I,$$

so the 'x^+' notation is a device that saves the enumeration of many different cases.

Now we can address the question of finding the least common multiple of all of the $\delta_{i,j}$'s in (4.4.8). For each s, a common multiple of all of the *falling* factorials that appear there will be the one whose first argument is largest, i.e.,

$$\mathrm{ff}((a_s)^+ J + (b_s)^+ I, a_s n + b_s k + c_s),$$

and a common multiple of all of the *rising* factorials that appear there will similarly be

$$\mathrm{rf}((-u_s)^+ J + (-v_s)^+ I, u_s n + v_s k + w_s).$$

Consequently the least common denominator of the expression (4.4.8), when that expression is thought of as a rational function of k, with n as a parameter, surely divides $P(n,k)$ times

$$\prod_{s=1}^{uu} \mathrm{ff}((a_s)^+ J + (b_s)^+ I, a_s n + b_s k + c_s) \prod_{s=1}^{vv} \mathrm{rf}((-u_s)^+ J + (-v_s)^+ I, u_s n + v_s k + w_s).$$
(4.4.9)

Therefore we can clear (4.4.8) of fractions if we multiply it through by (4.4.9). The result of multiplying (4.4.8) through by (4.4.9) will be the *polynomial in k*

$$\sum_{0 \leq i \leq I; 0 \leq j \leq J} a_{i,j}(n)\nu_{i,j}(n,k)\frac{\Delta}{\delta_{i,j}(n,k)},$$
(4.4.10)

in which Δ is the common denominator in (4.4.9).

In order to prove the theorem we must show that if I and J are large enough, then the system of linear equations in the unknown $a_{i,j}$'s that one obtains by equating to zero the coefficient of every power of k that appears in (4.4.10) actually has a nontrivial solution. This will surely happen if the number of unknowns exceeds the number of equations, and we claim that if I, J are large enough then this is exactly what happens.

Indeed, the number of unknown $a_{i,j}$'s is obviously $(I + 1)(J + 1)$. The number of equations that they must satisfy is the number of different powers of k that appear in (4.4.10). We claim that the number of different powers of k that appear there grows only linearly with I and J, that is, as $c_1 I + c_2 J + c_3$, where the c's are independent of I, J. This claim would be sufficient to prove the theorem because then the number of unknowns would grow like IJ, for large I and J, whereas the number of equations would grow only as $c_1 I + c_2 J + c_3$. Hence for large enough I, J the latter would be less than the former.

Since the degree in k of each rising factorial and each falling factorial that appears in (4.4.10) grows linearly with I, J, and there are only a fixed number of each of them, the degrees in k of all of the ν's, δ's and Δ grow linearly with I, J. Hence the claim is clearly true, and the proof of the main theorem is complete. A more detailed argument, which we omit here, shows that the values I^* and J^* that are in the statement of the theorem are already sufficiently large. $\square$

The above proof of the Fundamental Theorem is taken from Wilf and Zeilberger [WZ92a]. The theorem was proved earlier, in only slightly restricted generality, in the case of one summation variable, by Verbaeten [Verb74]. In [WZ92a] the theorem is stated and proved also for several summation variables, and for q and multi-q identities, in all cases with explicit *a priori* bounds for the order of the recurrence.

In some cases even more stupefying things are possible. Suppose $\sum_k F(n, k) = 1$ is an identity of the type that we have been considering. Then by the Fundamental Theorem, there exists an integer n_0 with the following property: suppose that we have *numerically* verified that the claimed identity is correct for $n = 0, 1, \ldots, n_0$. Then the identity is thereby *proved* to be true in general.

What is the integer n_0 that validates such a proof by computation? Once we have found the recurrence that the left side, $f(n)$, satisfies, suppose that recurrence turns out to be of order J. A function that satisfies a recurrence of order J, and that is equal to 1 for J consecutive values of n, has *not* been thereby proved to be 1 for all n.

The reason is that in the recurrence relation for f, say

$$a_0(n)f(n) + a_1(n)f(n-1) + \cdots + a_J(n)f(n-J) = 0$$

the coefficient $a_0(n)$ might vanish for some large values of n, and then we would not be able to solve for $f(n)$ from its predecessors and sustain the induction.

For example, if a certain sum $f(n)$ satisfies

$$(n - 100)f(n) - nf(n-1) + 100f(n-2) = 0$$

for all $n \geq 2$, and if we start checking that $f(n) = 1$, numerically from its definition as a sum, beginning with $n = 0$, then we will have to check up to $n = 100$ before we can safely conclude that $f(n) = 1$ for all $n \geq 0$.

In general, suppose $\sum_{i=0}^{I} \sum_{j=0}^{J} b_{i,j} n^j f(n-i) = 0$ is a linear recurrence with polynomial coefficients, and suppose that $f(n) = 1$ satisfies this recurrence for at least $J + 1$ consecutive values of n. Then we have $\sum_{i=0}^{I} \sum_{j=0}^{J} b_{i,j} n^j = 0$ for those n. Think of these as a set of $J + 1$ linear homogeneous equations in $J + 1$ unknowns $x_j = \sum_{i=0}^{I} b_{i,j}$. Since the coefficient determinant of this system is the

Vandermondian $\{n^j\}_{0 \le j \le J; 0 \le n \le J}$, it is nonsingular, and therefore all of the x_j's must vanish. Thus $f(n) = 1$ is a solution for all n. If the highest coefficient in the recurrence, namely $\sum_{j=0}^{J} b_{0,j}$, is nonzero, then the solution in which $f(n)$ is identically 1 will be the unique solution of the recurrence that satisfies the initial conditions.

So a safe estimate for n_0 is, for instance, the order of the recurrence plus the size of the largest nonnegative zero of the leading coefficient of the recurrence plus the highest degree in n of any of the coefficient polynomials.

Remarkably, it is possible to estimate, *a priori*, the roots of the leading coefficient of the recurrence relation. This has been done by Lily Yen in her doctoral dissertation [Yen 93]. Together with the *a priori* bounds on the order of the recurrence relation that we have already discussed here, as well as *a priori* bounds on the degrees of the coefficient polynomials that occur in the recurrence, this means that we can estimate n_0 *a priori* also. Theoretically, then, we can prove identities just by checking enough numerical values! As things now stand, however, the *a priori* estimates of n_0 are extremely large, so the algorithm is impractical. It is an interesting research question to ask how small we can make this *a priori* estimate of n_0.

In more recent work [Yen97], however, Yen has shown that for q-identities the *a priori* estimates of n_0 can be spectacularly reduced, since in that case she proved that the leading coefficient of the recurrence relation cannot vanish. This opens the door to *proving q-identities by simply verifying that they are satisfied for some moderate finite number of values of n*. In the case of the Chu–Vandermonde identity (see page 186), for instance, she shows that we can prove that it is true for all n "just" by checking it for 2358 values of n. While this is not yet a practical-sized computation, it hints that such things may lie just ahead.

4.5   Multivariate and "q" generalizations

The Fundamental Theorem generalizes to multivariate sums, i.e., sums over several summation indices, and to q- and multi-q- sums. These results are in Wilf and Zeilberger [WZ92a], and we will give here only a summary of the principal results of that paper.

First, here is the generalization to r summation indices, in which we are trying to find recurrences that are satisfied by sums of the form

$$f_n(\mathbf{x}) = \sum_{k_1, \ldots, k_r} F(n, k_1, k_2, \ldots, k_r) x_1^{k_1} \cdots x_r^{k_r} \tag{4.5.1}$$

for integer n, where $r \ge 1$ and the summand F is a proper hypergeometric term.

The allowable form of a proper hypergeometric summand F in this case is

$$F(n, \mathbf{k}) = P(n, \mathbf{k}) \frac{\prod_{s=1}^{p}(a_s n + \mathbf{b}_s \cdot \mathbf{k} + c_s)!}{\prod_{s=1}^{q}(u_s n + \mathbf{v}_s \cdot \mathbf{k} + w_s)!} \mathbf{z^k}, \qquad (4.5.2)$$

where P is a polynomial, the a's, u's, $\mathbf{b}$'s and $\mathbf{v}$'s are integers that contain no additional parameters, and the c's and w's are integers that may involve unspecified parameters.

The form of the k-free recurrence relation that these F's satisfy is

$$\sum_{0 \le j \le J} \sum_{0 \le \mathbf{i} \le \mathbf{I}} \alpha(\mathbf{i}, j, n) F(n - j, \mathbf{k} - \mathbf{i}) = 0, \qquad (4.5.3)$$

where the α's are polynomials in n.

Now here is the r-variate Fundamental Theorem.

Theorem 4.5.1 *Every proper hypergeometric term in r variables satisfies a nontrivial k-free recurrence relation. Indeed, there exist $\mathbf{I}$, J and polynomials $\alpha(\mathbf{i}, j, n)$, not all zero, such that (4.5.3) holds at every point $(n_0, \mathbf{k}_0) \in \mathbb{Z}^{r+1}$ for which $F(n_0, \mathbf{k}_0) \ne 0$ and all of the values $F(n_0 - j, \mathbf{k}_0 - \mathbf{i})$ that occur in (4.5.3) are well defined. Furthermore, there is such a recurrence in which $J = J^*$, where*

$$J^* = \left\lfloor \frac{1}{r!} \left\{ \sum_{s=1}^{p} \sum_{r'=1}^{r} |(\mathbf{b}_s)_{r'}| + \sum_{s=1}^{q} \sum_{r'=1}^{r} |(\mathbf{v}_s)_{r'}| \right\}^r \right\rfloor. \qquad (4.5.4)$$

Similarly, the Fundamental Theorem can be generalized to q-sums and multisums. We present here the theorem only in the case of q-sums.

A q-proper hypergeometric term is of the form

$$F(n, k) = \frac{\prod_s Q(a_s n + b_s k, c_s)}{\prod_s Q(u_s n + v_s k, w_s)} q^{an^2 + bnk + ck^2 + dk + en} \xi^k, \qquad (4.5.5)$$

where

$$Q(m, c) = (1 - cq)(1 - cq^2) \cdots (1 - cq^m). \qquad (4.5.6)$$

Our hypotheses about the parameters, etc. will be as above, i.e., they are all absolute constants except possibly for the c's and the w's. As before, we seek I, J such that for some nontrivial α's we have

$$\sum_{i=0}^{I} \sum_{j=0}^{J} \alpha(i, j; n) F(n - j, k - i) = 0.$$

The result is as follows.

Theorem 4.5.2 *Let F be a q-hypergeometric term of the form (4.5.5). Then F satisfies a k-free recurrence whose order J is at most $\sum_s b_s^2 + \sum_s v_s^2 + 2|c|$.*

For the proofs of these theorems and many examples thereof, the reader is referred to [WZ92a], and to Section 6.5 of this book. One can go even further, and talk about multiple-summation/integration [WZ92a]

$$f_n(x) := \int \cdots \int \sum_{k1,\ldots,k_r} F(x,n;y_1,\ldots,y_s;k_1,\ldots,k_r) dy_1 \ldots dy_s$$

where F is a continuous/discrete analog of 'proper hypergeometric.' The program `TRIPLE_INTEGRAL.maple`, with its associated sample input file `inTRIPLE`, as well as the program `DOUBLE_SUM_SINGLE_INTEGRAL.maple` are Maple implementations of two special cases.

4.6   Exercises

1. Find an upper bound, in terms of p, on the order of the recurrence that is satisfied by the sum of the pth powers of all of the binomial coefficients of order n.

2. Let $F(n,k)$ be a proper hypergeometric term, of the form (4.4.1), but without the polynomial factor P. Define $A = \sum_s a_s$, $B = \sum_s b_s$, $U = \sum_s u_s$, $V = \sum_s v_s$. Show that [Wilf91] the upper bounds I^*, J^* that were found in Theorem 4.4.1 can be replaced by

$$J^* = \sum_s (-v_s)^+ + \sum_s b_s^+ + (V - B)^+ \tag{4.6.1}$$

$$I^* = J^* \left\{ \sum a_s^+ + \sum (-u_s)^+ + (U - A)^+ - 1 \right\} + 1. \tag{4.6.2}$$

Investigate circumstances under which this bound is superior to the one stated in Theorem 4.4.1.

3. Use Sister Celine's algorithm to evaluate each of the following binomial coefficient sums, in explicit closed form. In each case find a recurrence that is satisfied by the summand, then sum the recurrence over the range of the given summation to find a recurrence that is satisfied by the sum. Then solve that recurrence for the sum, either by inspection, or by being very clever, or, *in extremis*, by using algorithm Hyper of Chapter 8, page 156.

   (a) $\sum_k (-1)^k \binom{n}{k} \binom{2n-2k}{n+a}$

(b) $\sum_k \binom{x}{k}\binom{y}{n-k}$

(c) $\sum_k k\binom{2n+1}{2k+1}$

(d) $\sum_{k=0}^{n} \binom{n+k}{k}2^{-k}$ (Careful– watch the limits of the sum.)

(e) $\sum_k (-1)^k \binom{n-k}{k}2^{n-2k}$

Chapter 5

Gosper's Algorithm

5.1 Introduction

Gosper's algorithm is one of the landmarks in the history of computerization of the problem of closed form summation. It not only definitively answers the question for which it was developed, but it is also vital in the operation of the creative telescoping algorithm of Chapter 6 and the WZ algorithm of Chapter 7.

The question for which it was developed is quite analogous to the question of indefinite integration in finite terms, so let's take a moment to look at that problem. Suppose we are given an integral $H(x) = \int_a^x f(t)dt$, where f is, say, continuous, and we are trying to "do" it, i.e., we are struggling to find some simple form for the function $H(x)$.

Certainly we are all finished if we can find a simple-looking function F ("antiderivative") such that $F' = f$, for then our answer is just that $H(x) = F(x) - F(a)$.

We remark at once that there is no question at all about the existence of an antiderivative. There always is one. In fact H itself is such a function! But that is totally unhelpful, because we are looking for an answer in "simple form." So the definition of simple form is vitally important. In this integration problem typically one defines certain elementary functions, such as polynomials, trigonometric functions, and so forth, along with a few elementary operations, such as addition and multiplication, extraction of roots, etc., and then one defines "simple form" to be the form of any function that is obtainable from the elementary functions by a finite sequence of operations.

With that kind of a setup, the integration problem is very difficult, and has been settled completely only fairly recently [Risc70].

Now consider the question of indefinite *summation* in closed form. Instead of an integral, we are looking at a sum

$$s_n = \sum_{k=0}^{n-1} t_k, \qquad (5.1.1)$$

where t_k is a hypergeometric term that does not depend on n, i.e., the consecutive-term ratio

$$r(k) = \frac{t_{k+1}}{t_k} \qquad (5.1.2)$$

is a rational function of k. We would like to express s_n in closed form,[1] that is, without using the summation sign, if possible.

We note that s_n plays the role of an antiderivative here. Instead of its derivative being the integrand, its difference is the summand. That is, $s_{n+1} - s_n = t_n$. Hence, just as in the integration problem, we are led to inquire if, *given a hypergeometric term t_n, there exists a hypergeometric term z_n, say, such that*

$$z_{n+1} - z_n = t_n. \qquad (5.1.3)$$

If we can somehow find such a function z_n then we will indeed have expressed the sum (5.1.1) in the simple form of a single hypergeometric term plus a constant. Conversely, any solution z_n of (5.1.3) will have the form

$$z_n = z_{n-1} + t_{n-1} = z_{n-2} + t_{n-2} + t_{n-1} = \ldots = z_0 + \sum_{k=0}^{n-1} t_k = s_n + c,$$

where $c = z_0$ is a constant.

Gosper's algorithm [Gosp78] answers the following question: **Given a hypergeometric term t_n, is there a hypergeometric term z_n satisfying (5.1.3)?**

If the answer is affirmative, then s_n can be expressed as a hypergeometric term plus a constant, and the algorithm outputs such a term. In this case we will say that t_n is *Gosper-summable*. On the other hand, if Gosper's algorithm returns a negative answer, then that *proves* that (5.1.3) has no hypergeometric solution.

R. W. Gosper, Jr., discovered his algorithm in conjunction with his work on the development of one of the first symbolic algebra programs, Macsyma. Because of his algorithm, Macsyma had a seemingly uncanny ability to find simple formulas for sums of the type (5.1.1).

In this chapter, we will use $\mathbb{N}$ to denote the set of all nonnegative integers, $\mathbb{N} = \{0, 1, 2, \ldots\}$. If $p(n)$ is a nonzero polynomial we will denote its leading coefficient by $\operatorname{lc} p(n)$. The degree of the zero polynomial will be taken to be $-\infty$.

---

[1]See page 145.

We will assume that all arithmetic operations take place in some field K of characteristic 0. In the examples, it will be the case that $K = \mathbb{Q}$, the field of rational numbers, or $K = \mathbb{Q}(x_1, x_2, \ldots, x_k)$ where $x_1, x_2, \ldots, x_k$ are algebraically independent over $\mathbb{Q}$, or $K = \mathbb{Q}(\alpha)$ where α is algebraic over $\mathbb{Q}$. Our rational functions have their coefficients in K, therefore we sometimes call K the *coefficient field*. A sequence t_n with elements in K is then a *hypergeometric term over K* if there are polynomials $p(n)$, $q(n)$ from $K[n]$ such that $p(n)t_{n+1} = q(n)t_n$ for all $n \in \mathbb{N}$, i.e., if t_n satisfies a first order linear homogeneous recurrence whose coefficients are polynomials in $K[n]$. For more information on our algebraic framework, see Section 8.2.

5.2 Hypergeometrics to rationals to polynomials

If z_n is a hypergeometric term that satisfies (5.1.3) then the ratio

$$\frac{z_n}{t_n} = \frac{z_n}{z_{n+1} - z_n} = \frac{1}{\frac{z_{n+1}}{z_n} - 1} \tag{5.2.1}$$

is clearly a rational function of n. So let

$$z_n = y(n)t_n,$$

where $y(n)$ is an (as yet unknown) rational function of n. Substituting $y(n)t_n$ for z_n in (5.1.3) reveals that $y(n)$ satisfies

$$r(n)y(n+1) - y(n) = 1, \tag{5.2.2}$$

where $r(n)$ is as in (5.1.2). This is a first-order linear recurrence relation with rational coefficients and constant right hand side. Thus we have reduced the problem of finding *hypergeometric solutions* of (5.1.3) to the problem of finding *rational solutions* of (5.2.2).

Later, in Chapter 8, we will see how to find rational (and hypergeometric) solutions of linear recurrences with rational coefficients, of *any* order. But in this special case, Gosper found an ingenious way to reduce the problem further to that of finding *polynomial solutions* of yet another first-order recurrence.

Assume that we can write

$$r(n) = \frac{a(n)}{b(n)} \frac{c(n+1)}{c(n)}, \tag{5.2.3}$$

where $a(n)$, $b(n)$, $c(n)$ are polynomials in n, and

$$\gcd(a(n), b(n+h)) = 1, \quad \text{for all nonnegative integers } h. \tag{5.2.4}$$

We will see in the next section that a factorization of this type exists for every rational function, and we will also give an algorithm to find it. Following Gosper's advice, we look for a nonzero rational solution of (5.2.2) in the form

$$y(n) = \frac{b(n-1)x(n)}{c(n)},\tag{5.2.5}$$

where $x(n)$ is an unknown rational function of n. Substitution of (5.2.3) and (5.2.5) into (5.2.2) shows that $x(n)$ satisfies

$$a(n)x(n+1) - b(n-1)x(n) = c(n).\tag{5.2.6}$$

And now a miracle happens.[2]

Theorem 5.2.1 *[Gosp78] Let $a(n)$, $b(n)$, $c(n)$ be polynomials in n such that equation (5.2.4) holds. If $x(n)$ is a rational function of n satisfying (5.2.6), then $x(n)$ is a polynomial in n.*

Proof. Let $x(n) = f(n)/g(n)$, where $f(n)$ and $g(n)$ are relatively prime polynomials in n. Then (5.2.6) can be rewritten as

$$a(n)f(n+1)g(n) - b(n-1)f(n)g(n+1) = c(n)g(n)g(n+1).\tag{5.2.7}$$

Suppose that the conclusion of the theorem is false. Then $g(n)$ is a non-constant polynomial. Let N be the largest integer such that $\gcd(g(n), g(n+N))$ is a non-constant polynomial; note that $N \geq 0$. Let $u(n)$ be a non-constant irreducible common divisor of $g(n)$ and $g(n+N)$. Since $u(n-N)$ divides $g(n)$ it follows from (5.2.7) that

$$u(n-N) \mid b(n-1)f(n)g(n+1).$$

Now $u(n-N)$ does not divide $f(n)$ since it divides $g(n)$, which is relatively prime to $f(n)$ by assumption. It also does not divide $g(n+1)$, or else $u(n)$ would be a non-constant common factor of $g(n)$ and $g(n+N+1)$, contrary to our choice of N. Therefore $u(n-N) \mid b(n-1)$ and hence $u(n+1) \mid b(n+N)$.

Similarly, it follows from (5.2.7) that

$$u(n+1) \mid a(n)f(n+1)g(n).$$

Again, $u(n+1)$ does not divide $f(n+1)$ by assumption. It also does not divide $g(n)$, or else $u(n)$ would be a non-constant common factor of $g(n-1)$ and $g(n+N)$, contrary to our choice of N. Hence $u(n+1) \mid a(n)$. But then, by the previous

---

[2]But see Exercise 10.

paragraph, $u(n+1)$ is a non-constant common factor of $a(n)$ and $b(n+N)$, contrary to (5.2.4).

This contradiction shows that $g(n)$ is constant, and so $x(n)$ is a polynomial in n. □

Finding hypergeometric solutions of (5.1.3) is therefore equivalent to finding polynomial solutions of (5.2.6). The correspondence between them is that if $x(n)$ is a nonzero polynomial solution of (5.2.6) then

$$z_n = \frac{b(n-1)x(n)}{c(n)}t_n$$

is a hypergeometric solution of (5.1.3), and vice versa. The question of how to find polynomial solutions of (5.2.6), if they exist, or to prove that there are none, if they don't exist, is the subject of Section 5.4 below.

Here, then, is an outline of Gosper's algorithm.

Gosper's Algorithm

INPUT: A hypergeometric term t_n.
OUTPUT: A hypergeometric term z_n satisfying (5.1.3), if one exists; $\sum_{k=0}^{n-1} t_k$, otherwise.

1. Form the ratio $r(n) = t_{n+1}/t_n$ which is a rational function of n.

2. Write $r(n) = \frac{a(n)}{b(n)}\frac{c(n+1)}{c(n)}$ where $a(n)$, $b(n)$, $c(n)$ are polynomials satisfying (5.2.4).

3. Find a nonzero polynomial solution $x(n)$ of (5.2.6), if one exists; otherwise return $\sum_{k=0}^{n-1} t_k$ and stop.

4. Return $\frac{b(n-1)x(n)}{c(n)}t_n$ and stop.

Once we have z_n, the sum that we are looking for is $s_n = z_n - z_0$. The lower summation bound need not be 0; for example, it may happen that the summand in (5.1.1) is undefined for certain integer values of k, and then we will want to start the summation at some large enough value to skip over all the singularities. If the lower summation bound is k_0, then everything goes through as before, except that now $s_n = z_n - z_{k_0}$.

Example 5.2.1. Let

$$S_n = \sum_{k=0}^{n} (4k+1) \frac{k!}{(2k+1)!} \,.$$

Can this sum be expressed in closed form? We recognize at a glance that the summand

$$t_n = (4n+1) \frac{n!}{(2n+1)!}$$

is a hypergeometric term. We will use Gosper's algorithm to see if S_n can be expressed as a hypergeometric term plus a constant. The upper summation bound is n rather than $n-1$, so let $s_n = S_{n-1}$. The term ratio

$$r(n) = \frac{t_{n+1}}{t_n} = \frac{(4n+5)\frac{(n+1)!}{(2n+3)!}}{(4n+1)\frac{n!}{(2n+1)!}} = \frac{4n+5}{2(4n+1)(2n+3)}$$

is rational in n as expected. The choice

$$a(n) = 1, \ b(n) = 2(2n+3), \ c(n) = 4n+1$$

clearly satisfies (5.2.3) and (5.2.4). Equation (5.2.6) thus becomes

$$x(n+1) - 2(2n+1)x(n) = 4n+1 \,. \tag{5.2.8}$$

Does it have any nonzero polynomial solution? We might start looking for polynomial solutions of degree $0, 1, 2, \dots$ until one is found. Here we "get lucky," since we find a solution right away, namely the constant polynomial $x(n) = -1$. Hence

$$z_n = \frac{-2(2n+1)}{4n+1}(4n+1)\frac{n!}{(2n+1)!} = -2\frac{n!}{(2n)!}$$

satisfies $z_{n+1} - z_n = t_n$. Finally, $s_n = z_n - z_0 = 2 - 2n!/(2n)!$, so the closed form we were looking for is

$$S_n = s_{n+1} = 2 - \frac{n!}{(2n+1)!} \,.$$

Notice that S_n is not a hypergeometric term. It is, however, the sum of two such terms, one of them constant.   □

This example was so simple that we were able to find the factorization (5.2.2) and a polynomial solution of (5.2.6) by inspection. It remains to show how to do Steps 2 and 3 in a systematic way.

5.3 The full algorithm: Step 2

In this section we show how to obtain the factorization (5.2.3) of a given rational function $r(n)$, subject to conditions (5.2.4), and derive some of its properties.

Let $r(n) = f(n)/g(n)$, where $f(n)$ and $g(n)$ are relatively prime polynomials. If $\gcd(f(n), g(n + h)) = 1$ for all nonnegative integers h, we can take $a(n) = f(n)$, $b(n) = g(n)$, $c(n) = 1$, and we would have the desired factorization at once. Otherwise let $u(n)$ be a non-constant common factor of $f(n)$ and $g(n + h)$, for some nonnegative integer h. The idea is to "divide out" such factors from the prospective $a(n)$ resp. $b(n)$ and to incorporate them into $c(n)$. More precisely, let $f(n) = \bar{f}(n)u(n)$ and $g(n) = \bar{g}(n)u(n - h)$. Then

$$r(n) = \frac{f(n)}{g(n)} = \frac{\bar{f}(n)}{\bar{g}(n)} \frac{u(n)}{u(n - h)}.$$

The last fraction on the right can be converted into a product of fractions of the form $c(n + 1)/c(n)$ by multiplying its numerator and denominator by the missing "intermediate" shifted factors of $u(n)$:

$$\frac{u(n)}{u(n - h)} = \frac{u(n)u(n - 1)u(n - 2) \cdots u(n - h + 1)}{u(n - 1)u(n - 2) \cdots u(n - h + 1)u(n - h)}.$$

Now we repeat this procedure with $\bar{f}$ and $\bar{g}$ in place of f and g. In a finite number of steps we will obtain the desired factorization (5.2.3).

How do we know when (5.2.4) is satisfied, or if it isn't, how do we find the values of h that violate it? One way is to use polynomial resultants.[3] Let $R(h)$ denote the resultant of $f(n)$ and $g(n + h)$, regarded as polynomials in n. Then $R(h)$ is a polynomial in h with the property that $R(\alpha) = 0$ if and only if $\gcd(f(n), g(n + \alpha))$ is not a constant polynomial. Therefore the values of h that violate (5.2.4) are precisely the nonnegative integer zeros of $R(h)$.

To speed up the computation of the resultant, we can replace f and g in the definition of $R(h)$ by $f/\gcd(f, f')$ and $g/\gcd(g, g')$, respectively. This is permitted by virtue of the fact that $f/\gcd(f, f')$ and f have the same sets of irreducible monic factors, and so do $g/\gcd(g, g')$ and g.

How can we find integer roots of a polynomial $R(h)$? The answer to this depends on the field K from which R's coefficients come. If $K = \mathbf{Q}$ this is easy, albeit possibly tedious: We can clear denominators in $R(h) = 0$ obtaining a new equation $S(h) = 0$, where S is a polynomial with integer coefficients and the same roots as R. If necessary, we cancel a power of h from this equation so that $S(0) \neq 0$. Now every integer root u of S divides the constant term of S since it divides all the other ones.

---

[3]The resultant of two polynomials f, g, is the product of the values of g at the zeros of f.

Gosper's Algorithm (Step 2)

Step 2.1. Let $r(n) = Z\frac{f(n)}{g(n)}$ where f, g are monic relatively prime
polynomials, and Z is a constant;
$R(h) := \text{Resultant}_n(f(n), g(n + h))$;
Let $S = \{h_1, h_2, \ldots, h_N\}$ be the set of nonnegative integer
zeros of $R(h)$ ($N \geq 0$, $0 \leq h_1 < h_2 < \ldots < h_N$).

Step 2.2. $p_0(n) := f(n)$; $q_0(n) := g(n)$;
for $j = 1, 2, \ldots, N$ do
$s_j(n) := \gcd(p_{j-1}(n), q_{j-1}(n + h_j))$;
$p_j(n) := p_{j-1}(n)/s_j(n)$;
$q_j(n) := q_{j-1}(n)/s_j(n - h_j)$.
$a(n) := Zp_N(n)$;
$b(n) := q_N(n)$;
$c(n) := \prod_{i=1}^{N} \prod_{j=1}^{h_i} s_i(n - j)$. $\square$

Thus a simple algorithm to find all integer roots of R, in the case $K = \mathbf{Q}$, is[4] to
check all divisors of the constant term of S.

More generally, if $K = k(\alpha)$ and A_k is an algorithm for finding integer roots
of polynomials with coefficients in k, then the corresponding algorithm A_K can be
obtained as follows: Let $R \in K[x]$. Since the elements of K are rational functions
of α we can write $R(h) = \sum_{i=0}^{s} p_i(\alpha)h^i/r(\alpha) = \sum_{j=0}^{t} q_j(h)\alpha^j/r(\alpha)$ where $p_i, q_j, r \in$
$k[x]$. Let $R(u) = 0$ for some $u \in k$. Then $\sum_{j=0}^{t} q_j(u)\alpha^j = 0$. If α is transcendental
over k it follows that $q_j(u) = 0$ for $j = 0, 1, \ldots, t$. If α is algebraic over k of degree d
then each p_i is of degree less than d, hence $t \leq d-1$. Again it follows that $q_j(u) = 0$
for $j = 0, 1, \ldots, t$. In either case, A_K consists of applying A_k to each of $q_j(u) = 0$,
for $j = 0, 1, \ldots, t$, and taking the intersection of the sets that are obtained.

Example 5.3.1. Let $R(h) = \sqrt{2}\,h^2 - 2(\sqrt{2} + 1)h + 4$. Here $K = \mathbf{Q}[\sqrt{2}]$. Rewrite
R as a polynomial in $\sqrt{2}$: $R(h) = h(h - 2)\sqrt{2} - 2(h - 2)$. One coefficient has roots
0 and 2, and the other has root 2, hence $h = 2$ is the only integer zero of $R(h)$. $\square$

We have to show that $a(n)$, $b(n)$ and $c(n)$ produced by this procedure for doing Step 2 of Gosper's algorithm satisfy conditions (5.2.3) and (5.2.4). A short

---

[4]A more efficient algorithm using p-adic methods is given in [Loos83].

computation verifies the former:

$$
\frac{a(n)}{b(n)}\frac{c(n+1)}{c(n)} = Z\frac{p_N(n)}{q_N(n)}\prod_{i=1}^{N}\prod_{j=1}^{h_i}\frac{s_i(n+1-j)}{s_i(n-j)}
$$

$$
= Z\frac{p_0(n)}{\prod_{i=1}^{N}s_i(n)}\frac{\prod_{i=1}^{N}s_i(n-h_i)}{q_0(n)}\prod_{i=1}^{N}\frac{s_i(n)}{s_i(n-h_i)}
$$

$$
= Z\frac{p_0(n)}{q_0(n)} = Z\frac{f(n)}{g(n)} = r(n).
$$

To verify the latter, note that by definition of p_j, q_j, and s_j,

$$
\gcd(p_k(n), q_k(n+h_k)) = \gcd\left(\frac{p_{k-1}(n)}{s_k(n)}, \frac{q_{k-1}(n+h_k)}{s_k(n)}\right) = 1 \qquad (5.3.1)
$$

for all k such that $1 \le k \le N$. In fact, more is true. We use the notation from Step 2 of Gosper's algorithm, and define additionally $h_{N+1} := +\infty$.

Proposition 5.3.1 *Let $0 \le k \le i, j \le N$, $h \in \mathbb{N}$ and $h < h_{k+1}$. Then*

$$
\gcd(p_i(n), q_j(n+h)) = 1. \qquad (5.3.2)
$$

Proof. Since $p_i(n) \mid f(n)$ and $q_j(n) \mid g(n)$, it follows that $\gcd(p_i(n), q_j(n+h))$ divides $\gcd(f(n), g(n+h))$, for any h. If $h \in \mathbb{N}$ but $h \notin S$ then $R(h) \ne 0$, hence $\gcd(f(n), g(n+h)) = 1$. This proves the assertion when $h \notin S$.

To prove it when $h \in S$, we use induction on k. Recall that S is sorted so that $h_1 < h_2 < \ldots < h_N$.

$k = 0$: In this case there is nothing to prove since there is no $h \in S$ such that $h < h_1$.

$k > 0$: Assume that the assertion holds for all $h < h_k$. It remains to show that it holds for $h = h_k$. Since $p_i(n) \mid p_k(n)$ and $q_j(n) \mid q_k(n)$ it follows that $\gcd(p_i(n), q_j(n+h_k))$ divides $\gcd(p_k(n), q_k(n+h_k))$. By (5.3.1) the latter gcd is 1, completing the proof. $\qquad \square$

Setting $i = j = k = N$ in (5.3.2) we see that $\gcd(a(n), b(n+h)) = 1$ for all $h \in \mathbb{N}$, proving (5.2.4).

It is easy to see that (5.2.4) will be satisfied by the output of Step 2, regardless of the order in which the members of S are considered. But if they are considered in increasing order then we claim that the resulting $c(n)$ will have the lowest possible degree among all factorizations (5.2.3) which satisfy (5.2.4). This is important since $c(n)$ is the denominator of the unknown rational function $y(n)$, and thus the size of the linear system resulting from (5.2.6) will be the least possible as well.

Theorem 5.3.1 *Let K be a field of characteristic zero and $r \in K[n]$ a nonzero rational function. Then there exist polynomials $a, b, c \in K[n]$ such that b, c are monic and*

$$r(n) = \frac{a(n)}{b(n)} \frac{c(n+1)}{c(n)}, \tag{5.3.3}$$

where

(i) $\gcd(a(n), b(n+h)) = 1$ *for every nonnegative integer h,*

(ii) $\gcd(a(n), c(n)) = 1$,

(iii) $\gcd(b(n), c(n+1)) = 1$.

Such polynomials are constructed by Step 2 of Gosper's algorithm.

Proof. Let $a(n), b(n)$ and $c(n)$ be the polynomials produced by Step 2 of Gosper's algorithm. We have already shown (in the discussion preceding the statement of the theorem) that (5.3.3) and **(i)** are satisfied.

(ii): If $a(n)$ and $c(n)$ have a non-constant common factor then so do $p_N(n)$ and $s_i(n-j)$, for some i and j such that $1 \le i \le N$ and $1 \le j \le h_i$. Since by definition $q_{i-1}(n+h_i-j) = q_i(n+h_i-j)s_i(n-j)$, it follows that $p_N(n)$ and $q_{i-1}(n+h_i-j)$ have such a common factor, too. Since $h_i - j < h_i$, this contradicts Proposition 5.3.1. Hence $a(n)$ and $c(n)$ are relatively prime.

(iii): If $b(n)$ and $c(n+1)$ have a non-constant common factor then so do $q_N(n)$ and $s_i(n-j)$, for some i and j such that $1 \le i \le N$ and $0 \le j \le h_i - 1$. Since by definition $p_{i-1}(n-j) = p_i(n-j)s_i(n-j)$, it follows that $p_{i-1}(n)$ and $q_N(n+j)$ have such a common factor, too. Since $j < h_i$, this contradicts Proposition 5.3.1. Hence $b(n)$ and $c(n+1)$ are relatively prime. $\quad\square$

The following lemma will be useful more than once.

Lemma 5.3.1 *Let K be a field of characteristic zero. Let a, b, c, A, B, $C \in K[n]$ be polynomials such that $\gcd(a(n), c(n)) = \gcd(b(n), c(n+1)) = \gcd(A(n), B(n+h)) = 1$, for all nonnegative integers h. If*

$$\frac{a(n)}{b(n)} \frac{c(n+1)}{c(n)} = \frac{A(n)}{B(n)} \frac{C(n+1)}{C(n)}, \tag{5.3.4}$$

then $c(n)$ divides $C(n)$.

Proof. Let

$$g(n) = \gcd(c(n), C(n)), \tag{5.3.5}$$

$$d(n) = c(n)/g(n), \tag{5.3.6}$$

$$D(n) = C(n)/g(n). \tag{5.3.7}$$

Then $\gcd(d(n), D(n)) = \gcd(a(n), d(n)) = \gcd(b(n), d(n+1)) = 1$. Rewrite (5.3.4) as $A(n)b(n)c(n)C(n+1) = a(n)B(n)C(n)c(n+1)$ and cancel $g(n)g(n+1)$ on both sides. The result $A(n)b(n)d(n)D(n+1) = a(n)B(n)D(n)d(n+1)$ shows that

$$d(n) \mid B(n)d(n+1)$$
$$d(n+1) \mid A(n)d(n).$$

Using these two relations repeatedly, one finds that

$$d(n) \mid B(n)B(n+1)\cdots B(n+k-1)d(n+k),$$
$$d(n) \mid A(n-1)A(n-2)\cdots A(n-k)d(n-k),$$

for all $k \in \mathbb{N}$. Since K has characteristic zero, $\gcd(d(n), d(n+k)) = \gcd(d(n), d(n-k)) = 1$ for all large enough k. It follows that $d(n)$ divides both $B(n)B(n+1)\cdots B(n+k-1)$ and $A(n-1)A(n-2)\cdots A(n-k)$ for all large enough k. But these two polynomials are relatively prime by assumption, so $d(n)$ is a constant. Hence $c(n)$ divides $C(n)$, by (5.3.6) and (5.3.5). $\qquad\square$

Corollary 5.3.1 *Let $r(n)$ be a rational function. The factorization (5.3.3) described in Theorem 5.3.1 is unique.*

Proof. Assume that

$$r(n) = \frac{a(n)}{b(n)}\frac{c(n+1)}{c(n)} = \frac{A(n)}{B(n)}\frac{C(n+1)}{C(n)}$$

where polynomials a, b, c, A, B, C satisfy properties (i), (ii), (iii) of Theorem 5.3.1 and b, c, B, C are monic. By Lemma 5.3.1, $c(n)$ divides $C(n)$ and vice versa. As they are both monic, $c(n) = C(n)$. Therefore $A(n)b(n) = a(n)B(n)$. By property *(i)* of Theorem 5.3.1, $b(n)$ divides $B(n)$ and vice versa, so $b(n) = B(n)$ since they are monic. Hence $a(n) = A(n)$ as well. $\qquad\square$

This shows that the factorization described in Theorem 5.3.1 and computed by Step 2 of Gosper's algorithm is in fact a *canonical form* for rational functions.

Corollary 5.3.2 *Among all triples $a(n), b(n), c(n)$ satisfying (5.2.3) and (5.2.4), the one constructed in Step 2 of Gosper's algorithm has $c(n)$ of least degree.*

Proof. Let $A(n)$, $B(n)$, $C(n)$ satisfy (5.2.3) and (5.2.4). By Theorem 5.3.1, the $a(n), b(n), c(n)$ produced in Step 2 of Gosper's algorithm satisfy properties (ii) and (iii) of that theorem. Then it follows from Lemma 5.3.1 that $c(n)$ divides $C(n)$. $\square$

Example 5.3.2. Let $r(n) = (n+3)/((n(n+1))$. Then Step 2 of Gosper's algorithm yields $a(n) = 1$, $b(n) = n$, $c(n) = (n+1)(n+2)$. Note that (5.2.3) and

(5.2.4) will be also satisfied by, for example, $a(n) = n - k$, $b(n) = n(n - k + 1)$, $c(n) = (n + 1)(n + 2)(n - k)$, where k is any positive integer. However, here properties (ii) and (iii) of Theorem 5.3.1 are violated.   □

We conclude this section by discussing an alternative way of finding the set S in Step 2, which does not require computation of resultants. It consists of *factoring* polynomials $f(n)$ and $g(n)$ into irreducible factors, then finding all pairs $u(n), v(n)$ of irreducible factors of $f(n)$ resp. $g(n)$ such that

$$v(n) = u(n - h) \tag{5.3.8}$$

for some $h \in \mathbb{N}$. Namely, if $f(n)$ and $g(n + h)$ have a non-constant common factor, they also have a monic irreducible such factor, say $u(n)$, hence $g(n)$ has an irreducible factor of the form (5.3.8). An obvious necessary condition for (5.3.8) to hold is that u and v be of the same degree. If $u(n) = n^d + An^{d-1} + O(n^{d-2})$ and $v(n) = n^d + Bn^{d-1} + O(n^{d-2})$, then by comparing terms of order $d - 1$ in (5.3.8) we see that $h = (A - B)/d$ is the only choice for the value of the shift. It remains to check if $h \in \mathbb{N}$ and if (5.3.8) holds for this h.

In practice, $f(n)$ and $g(n)$ are usually already factored into linear factors because we get them from products of factorials and binomials. Then the resultant-based method and the factorization-based method come down to the same thing, since resultants are multiplicative in both arguments, and $\text{Resultant}_n(n + A, n + B + h) = h - (A - B)$. Even when f and g come unfactored it often seems a good idea to factor them first in order to speed up computation of the resultant — so why use resultants at all? On the other hand, the resultant-based method is more general since resultants can be computed in any field (by evaluating a certain determinant), whereas a generic polynomial factorization algorithm is not known. Ultimately, our choice of method will be based on availability and complexity of algorithms for computing resultants vs. polynomial factorizations over the coefficient field K.

5.4   The full algorithm: Step 3

Next we explain how to look for nonzero polynomial solutions of (5.2.6) in a systematic way. Assume that $x(n)$ is a polynomial that satisfies (5.2.6), with $\deg x(n) = d$. If we knew d, or at least had an upper bound for it, we could simply substitute a generic polynomial of degree d for $x(n)$ into (5.2.6), equate the coefficients of like powers of n, and solve the resulting equations for the unknown coefficients of $x(n)$. Note that these equations will be linear since (5.2.6) is linear in $x(n)$.

As it turns out, it is not difficult to obtain a finite set of candidates (at most two, in fact) for d. We distinguish two cases.

Case 1: $\deg a(n) \neq \deg b(n)$ or $\mathrm{lc}\, a(n) \neq \mathrm{lc}\, b(n)$.

The leading terms on the left of (5.2.6) do not cancel. Hence the degree of the left hand side of (5.2.6) is $d + \max\{\deg a(n), \deg b(n)\}$. Since the degree of the right hand side is $\deg c(n)$, it follows that

$$d = \deg c(n) - \max\{\deg a(n), \deg b(n)\}$$

is the only candidate for the degree of a nonzero polynomial solution of (5.2.6).

Case 2: $\deg a(n) = \deg b(n)$ and $\mathrm{lc}\, a(n) = \mathrm{lc}\, b(n) = \lambda$.

The leading terms on the left of (5.2.6) cancel. Again there are two cases to consider.

   (2a) The terms of second-highest degree on the left of (5.2.6) do not cancel. Then the degree of the left hand side of (5.2.6) is $d + \deg a(n) - 1$, thus

$$d = \deg c(n) - \deg a(n) + 1.$$

   (2b) The terms of second-highest degree on the left of (5.2.6) cancel. Let

$$a(n) = \lambda n^k + A n^{k-1} + O(n^{k-2}), \tag{5.4.1}$$

$$b(n-1) = \lambda n^k + B n^{k-1} + O(n^{k-2}), \tag{5.4.2}$$

$$x(n) = C_0 n^d + C_1 n^{d-1} + O(n^{d-2})$$

where $C_0 \neq 0$. Then, expanding the terms on the left of (5.2.6) successively, we find that

$$x(n+1) = C_0 n^d + (C_0 d + C_1) n^{d-1} + O(n^{d-2}),$$

$$a(n)x(n+1) = C_0 \lambda n^{k+d} + (\lambda(C_0 d + C_1) + A C_0) n^{k+d-1} + O(n^{k+d-2}),$$

$$b(n-1)x(n) = C_0 \lambda n^{k+d} + (B C_0 + \lambda C_1) n^{k+d-1} + O(n^{k+d-2}),$$

$$a(n)x(n+1) - b(n-1)x(n) = C_0(\lambda d + A - B) n^{k+d-1} + O(n^{k+d-2}). \tag{5.4.3}$$

By assumption, the coefficient of n^{k+d-1} on the right hand side of (5.4.3) vanishes, therefore $C_0(\lambda d + A - B) = 0$. It follows that

$$d = \frac{B - A}{\lambda}.$$

   Thus in Case 2 the only possible degrees of nonzero polynomial solutions of (5.2.6) are $\deg c(n) - \deg a(n) + 1$ and $(B - A)/\lambda$, where A and B are defined by (5.4.1) and (5.4.2), respectively. Of course, only nonnegative integer candidates need be considered. When there are two candidates we can use the larger of the two as an upper bound for the degree. Note that, in general, both Cases (2a) and (2b) can occur since equation (5.2.6) may in fact have nonzero polynomial solutions of two distinct degrees.

Example 5.4.1. Consider the sum $\sum_{k=1}^{n} 1/(k(k+1))$. Here $t_{n+1}/t_n = n/(n+2)$, hence $a(n) = n$, $b(n) = n+2$, $c(n) = 1$ and equation (5.2.6) is

$$nx(n+1) - (n+1)x(n) = 1.$$

Case 1 does not apply here. Case (2a) yields $d = 0$, and Case (2b) yields $d = 1$ as the only possible degrees of polynomial solutions. Indeed, the general solution of this equation is $x(n) = Cn - 1$, so there are solutions of degree 0 (when $C = 0$) and solutions of degree 1 (when $C \neq 0$). $\qquad\qquad\qquad\qquad\qquad\qquad\qquad$ $\square$

Gosper's Algorithm (Step 3)

Step 3.1. If $\deg a(n) \neq \deg b(n)$ or $\operatorname{lc} a(n) \neq \operatorname{lc} b(n)$ then
$\qquad\quad \mathcal{D} := \{\deg c(n) - \max\{\deg a(n), \deg b(n)\}\}$
$\quad$ else
$\qquad\quad$ let A and B be as in (5.4.1) and (5.4.2), respectively;
$\qquad\quad \mathcal{D} := \{\deg c(n) - \deg a(n) + 1,\ (B - A)/\operatorname{lc} a(n)\}$.
$\quad$ Let $\mathcal{D} := \mathcal{D} \cap \mathbb{N}$.
$\quad$ If $\mathcal{D} = \emptyset$ then return "no nonzero polynomial solution" and stop
$\quad$ else $d := \max \mathcal{D}$.

Step 3.2. Using the method of undetermined coefficients, find a nonzero
$\qquad\quad$ polynomial solution $x(n)$ of (5.2.6), of degree d or less.
$\qquad$ If none exists return "no nonzero polynomial solution" and stop.

$\square$

5.5   More examples

Example 5.5.1. Does the sum of the first $n + 1$ factorials

$$S_n = \sum_{k=0}^{n} k!$$

have a closed form? Here $t_n = n!$ and $r(n) = t_{n+1}/t_n = n + 1$, so we can take $a(n) = n+1$, $b(n) = c(n) = 1$. The equation (5.2.6) is

$$(n+1)x(n+1) - x(n) = 1,$$

and we are in Case 1 since $\deg a(n) \neq \deg b(n)$. The sole candidate for the degree of $x(n)$ is $\deg c(n) - \deg a(n) = -1$, so (5.2.6) has no nonzero polynomial solution in this case, proving that our sum cannot be written as a hypergeometric term plus a constant. $\qquad\qquad\qquad\qquad\qquad\qquad\qquad\qquad\qquad\qquad$ $\square$

Example 5.5.2. Modify the above sum so that it becomes

$$S_n = \sum_{k=1}^{n} kk!$$

and see what happens. Now $t_n = nn!$ and $r(n) = t_{n+1}/t_n = (n+1)^2/n$, hence $a(n) = n + 1$ and $b(n) = 1$ as before, but $c(n) = n$. The equation (5.2.6) is

$$(n + 1)x(n + 1) - x(n) = n, \qquad (5.5.1)$$

and we are in Case 1 again, but now the candidate for the degree of $x(n)$ is $\deg c(n) - \deg a(n) = 0$. Indeed, $x(n) = 1$ satisfies (5.5.1), thus $z_n = n!$ satisfies (5.1.3), $s_n = z_n - z_1 = n! - 1$, and $S_n = s_{n+1} = (n + 1)! - 1$. □

These two examples remind us again of integration where, e.g., $\int e^{x^2} dx$ is not an elementary function, while $\int xe^{x^2} dx = e^{x^2}/2 + C$ is.

Now we stop doing examples by hand and turn on the computer. After invoking Mathematica we read in the package gosper.m provided in the Mathematica programs that accompany this book (see Appendix A):

$$\text{In}[1] :=<< \text{gosper.m}$$

An easy example that we have already done by hand shows the syntax for doing a sum with given bounds:

$$\text{In}[2] := \text{GosperSum}[k\,k!,\ \{k, 0, n\}]$$

The answer agrees with our earlier result:

$$\text{Out}[2] = -1 + (1 + n)!$$

We can also require the indefinite sum by giving only the summation variable as the second argument:

$$\text{In}[3] := \text{GosperSum}[k\,k!,\ k]$$

What we get is a function $S(k)$ such that $S(k + 1) - S(k)$ equals $kk!$:

$$\text{Out}[3] = k!$$

When the summand is not Gosper-summable we get back the sum unchanged, except that it is now an ordinary Mathematica Sum:

$$\text{In}[4] := \text{GosperSum}[\text{Binomial}[n, k],\ \{k, 0, n\}]$$

$$\text{Out}[4] = \text{Sum}[\text{Binomial}[n, k],\ \{k, 0, n\}]$$

This answer means that the equation $z(k+1) - z(k) = \binom{n}{k}$ has no hypergeometric solution over the field $\mathbf{Q}(n)$. In other words, the "indefinite" sum $S(m) = \sum_{k=0}^{m} \binom{n}{k}$ is not expressible as a hypergeometric term over $\mathbf{Q}(n)$, plus a constant. (Note, however, that $S(n) = \sum_{k=0}^{n} \binom{n}{k} = 2^n$ *is* hypergeometric over $\mathbf{Q}$. Algorithms to evaluate such definite sums will be given in Chapter 6.)

A small change, the factor $(-1)^k$, makes the function $\binom{n}{k}$ Gosper-summable.

$$\text{In}[5] := \texttt{GosperSum}[(-1)\hat{\ }\texttt{k Binomial}[n, k], \{k, 0, n\}]$$

$$\text{Out}[5] = 0$$

Of course, this means that the algorithm will succeed with a general upper summation bound, too:

$$\text{In}[6] := \texttt{GosperSum}[(-1)\hat{\ }\texttt{k Binomial}[n, k], \{k, 0, m\}]$$

$$\text{Out}[6] = \frac{(-1)\hat{\ }\texttt{m}\,(-\texttt{m} + \texttt{n})\,\texttt{Binomial}[\texttt{n}, \texttt{m}]}{\texttt{n}}$$

Our next example is problem 10229 from the *American Mathematical Monthly* 99 (1992), p. 570. The summand contains an additional parameter m, hence the coefficient field is $\mathbf{Q}(m)$.

$$\text{In}[7] := \texttt{GosperSum}[\texttt{Binomial}[1/2, m - j + 1]\,\texttt{Binomial}[1/2, m + j], \{j, 1, p\}]$$

$$\text{Out}[7] = \frac{(1 + \texttt{m} - \texttt{p})\,\texttt{p}\,(-1 + 2\texttt{m} + 2\texttt{p})\,\texttt{Binomial}[1/2, 1 + \texttt{m} - \texttt{p}]\,\texttt{Binomial}[1/2, \texttt{m} + \texttt{p}]}{\texttt{m}\,(1 + 2\texttt{m})}$$

Here is another interesting example.

$$\text{In}[8] := \texttt{GosperSum}[\texttt{Binomial}[2k, k]/4\hat{\ }\texttt{k}, \{k, 0, n\}]$$

$$\text{Out}[8] = \frac{(1 + 2\texttt{n})\,\texttt{Binomial}[2\texttt{n}, \texttt{n}]}{4\hat{\ }\texttt{n}}$$

Let's see if this function is perhaps Gosper-summable again?

$$\text{In}[9] := \texttt{GosperSum}[\%, \{n, 0, n\}]$$

Yes indeed!

$$\text{Out}[9] = \frac{(1 + 2\texttt{n})\,(3 + 2\texttt{n})\,\texttt{Binomial}[2\texttt{n}, \texttt{n}]}{3 \quad 4\hat{\ }\texttt{n}}$$

Let's try this again:

$$\text{In}[10] := \texttt{GosperSum}[\%, \{n, 0, n\}]$$

$$\text{Out}[10] = \frac{(1 + 2\texttt{n})\,(3 + 2\texttt{n})\,(5 + 2\texttt{n})\,\texttt{Binomial}[2\texttt{n}, \texttt{n}]}{15 \quad 4\hat{\ }\texttt{n}}$$

And again:

$$\text{In}[11] := \text{GosperSum}[\%, \{n, 0, n\}]$$

$$\text{Out}[11] = \frac{(1 + 2n)\,(3 + 2n)\,(5 + 2n)\,(7 + 2n)\,\text{Binomial}[2n, n]}{105 \quad 4\char`\^n}$$

Now we can recognize the pattern. The numerical factors in the denominators are $1, 1 \times 3, 1 \times 3 \times 5, 1 \times 3 \times 5 \times 7$, so it looks as if

$$\sum_{n_s=0}^{n} \sum_{n_{s-1}=0}^{n_s} \cdots \sum_{n_1=0}^{n_2} \frac{\binom{2n_1}{n_1}}{4^{n_1}} = \frac{(2n + 2s - 1)!!}{(2n - 1)!!(2s - 1)!!} \frac{\binom{2n}{n}}{4^n} = \frac{\binom{2n+2s}{2s} \binom{2n}{n}}{\binom{n+s}{s} \; 4^n}, \qquad (5.5.2)$$

where the *double factorial* $n!!$ denotes the solution of the recurrence $a_n = na_{n-2}$ that satisfies $a_0 = a_1 = 1$. Our computer and Gosper's algorithm helped us guess this identity which contains an arbitrary number of summation signs. This is no proof, of course, but we can prove it by induction on s, using Gosper's algorithm again! The identity certainly holds for $s = 0$. Now let's assume that it holds when there are s summation signs present, and sum it once more. We will of course let Mathematica and Gosper's algorithm do it for us.

$$\text{In}[12] := \text{f}[\text{n\_}, \text{s\_}] := \text{Binomial}[2n + 2s, 2s]\,\text{Binomial}[2n, n]/\text{Binomial}[n + s, s]/4\char`\^n$$

First let's quickly check the base case:

$$\text{In}[13] := \text{f}[n, 0]$$

$$\text{Out}[13] = \frac{\text{Binomial}[2n, n]}{4\char`\^n}$$

And now for the induction step:

$$\text{In}[14] := \text{GosperSum}[\text{f}[n, s], \{n, 0, n\}]$$

$$\text{Out}[14] = \frac{(1 + 2n + 2s)\,\text{Binomial}[2n, n]\,\text{Binomial}[2n + 2s, 2s]}{4\char`\^n\,(1 + 2s)\,\text{Binomial}[n + s, s]}$$

$$\text{In}[15] := \% - \text{f}[n, s + 1] \; // \; \text{FactorialSimplify}$$

$$\text{Out}[15] = 0$$

This last zero means that if we take $f(n, s)$ and sum it again on n from 0 to n, what we get is exactly $f(n, s + 1)$, completing the proof.

There is another way of proving the induction step which does not need Gosper's algorithm, namely taking the difference of $f(n, s)$ w.r.t. n and showing that it is equal to $f(n, s - 1)$.

$$\text{In}[16] := (\mathtt{f}[\mathtt{n}, \mathtt{s}] - \mathtt{f}[\mathtt{n} - 1, \mathtt{s}]) - \mathtt{f}[\mathtt{n}, \mathtt{s} - 1] \mathbin{/\!/} \mathtt{FactorialSimplify}$$

$$\text{Out}[16] = 0$$

This means that $f(n, s)$ is correct to within an additive constant. To finish the proof, we have to show that $f(n, s)$ agrees with the left hand side of (5.5.2) at least for one value of n. Sure enough, for $n = 0$ both sides of (5.5.2) are equal to 1. For a generalization of this example, see Exercise 6 of this chapter.

If we want to find, instead, the solution $y(n)$ of Gosper's equation (5.2.2), then we can use the command $\mathtt{GosperFunction}$. For example, Gosper in his seminal paper [Gosp78] evaluates the sum

$$S_m = \sum_{n=1}^{m} \frac{\prod_{j=1}^{n-1} \left(bj^2 + cj + d\right)}{\prod_{j=1}^{n} \left(bj^2 + cj + e\right)}$$

assuming $d \neq e$. Here the consecutive-term ratio is

$$r(n) = \frac{bn^2 + cn + d}{b(n + 1)^2 + c(n + 1) + e},$$

and Gosper's algorithm

$$\text{In}[17] := \mathtt{GosperFunction}[(\mathtt{b\,n}\mathtt{\char94}2 + \mathtt{c\,n} + \mathtt{d})/(\mathtt{b}(\mathtt{n} + 1)\mathtt{\char94}2 + \mathtt{c}(\mathtt{n} + 1) + \mathtt{e}), \mathtt{n}]$$

finds that $y(n)$ is

$$\text{Out}[17] = \frac{\mathtt{e} + \mathtt{c\,n} + \mathtt{b\,n}\mathtt{\char94}2}{\mathtt{d} - \mathtt{e}}$$

over $\mathbb{Q}(b, c, d, e)$. Now $S_m = s_{m+1} = z_{m+1} - z_1$, where

$$z_m = y(m)t_m = \frac{1}{d - e} \frac{\prod_{j=1}^{m-1} \left(bj^2 + cj + d\right)}{\prod_{j=1}^{m-1} \left(bj^2 + cj + e\right)},$$

hence the final result is

$$S_m = \frac{1}{d - e} \left(\frac{\prod_{j=1}^{m} \left(bj^2 + cj + d\right)}{\prod_{j=1}^{m} \left(bj^2 + cj + e\right)} - 1 \right).$$

In Maple, Gosper's algorithm is one of the summation methods used by the built-in function sum:

```
> sum(4^k/binomial(2*k, k), k=0..n);
```

$$\begin{array}{c}
\text{n}\\
4 \ (2 \ \text{n} + 1)\\
4/3 \ \text{-----------------} \ + \ 1/3\\
\text{binomial}(2 \ \text{n} + 2, \ \text{n} + 1)
\end{array}$$

Here is an interesting example due to A. Giambruno and A. Regev. They proved an important result in the theory of polynomial identity algebras. However their result depends upon the hypothesis that

$$f(n) \neq g(n), \qquad \text{for } n \geq 5, \tag{5.5.3}$$

where

$$f(n) = \frac{(-1)^{n+1}(n^2 + 6n + 2)}{(2n + 1)(n + 2)} + \frac{(n + 1)!\, n!\, (2n^2 - 5n - 4)}{(2n + 1)!},$$

$$g(n) = (n + 1)!\, (n - 2)! \sum_{i=0}^{n-5} \frac{(-1)^i(n - i - 4)p(i, n)}{i!\, (2n - i - 3)!\, (i + 3)(i + 2)(n + 2)(2n - i - 2)},$$

and $p(i, n) = n^3 - 2in^2 - 3n^2 + i^2n + in - 4n + i^2 + 5i + 6$. It turns out that Gosper's algorithm succeeds on the sum in $g(n)$, so Maple can provide a closed form for the difference $f(n) - g(n)$.

```
> f := ((-1)^(n + 1))*((n^2 + 6*n + 2)/((2*n + 1)*(n + 2))) +
>       ((n + 1)!*n!*(2*n^2 - 5*n - 4))/((2*n + 1)!):
> g := (n + 1)!*(n - 2)!*
>       sum( (-1)^i*(n - i - 4)*(n^3 - 2*i*n^2 - 3*n^2 +
>            i^2*n + i*n - 4*n + i^2 + 5*i + 6)/(i!*
>            (2*n - i - 3)!*(i + 3)*(i + 2)*(n + 2)*(2*n - i - 2)),
>       i=0..n-5):
> g := expand(g):
> f := expand(f):
> factor(normal(expand(simplify(normal(f - g)))));
```

```
                          2              n
              (n + 1) (n  - 2 n + 2) (-1)
        - 3 ------------------------------
              (n + 2) (- 3 + 2 n) (2 n - 1)
```

This vanishes only for $n = -1, 1 \pm i$, proving (5.5.3).

5.6 Similarity among hypergeometric terms

The set of hypergeometric terms is closed under multiplication and reciprocation but not under addition. For example, while $n^2 + 1$ is a hypergeometric term, $2^n + 1$ isn't, although it is a nice and well behaved expression. In this section we answer the following:

Question 1. Given a hypergeometric term t_n, how can we decide if the sum $s_n = \sum_{k=0}^{n} t_k$ is expressible as a linear combination of several (but a *fixed number of*) hypergeometric terms? For example, since $k!$ is not Gosper-summable we know that the sum $\sum_{k=0}^{n} k!$ cannot be expressed as a hypergeometric term plus a constant; but could it be equal to a sum of two, or three, or any *fixed* (independent of n) number of hypergeometric terms?

Question 2. Given a linear combination c_n of hypergeometric terms, how can we decide if the sum $s_n = \sum_{k=0}^{n} c_k$ is expressible in the same form, that is, as a linear combination of hypergeometric terms? Note that such a combination may be Gosper-summable even though its individual terms are not. For example, take $t_{k+1} - t_k$ where t_k is a hypergeometric term which is not Gosper-summable.

In considering sums of hypergeometric terms, an important role is played by the relation of similarity.

Definition 5.6.1 *Two hypergeometric terms s_n and t_n are* similar *if their ratio is a rational function of n. In this case we write $s_n \sim t_n$.* $\square$

Similarity is obviously an equivalence relation in the set of all hypergeometric terms. One equivalence class, for example, consists of all rational functions.

Proposition 5.6.1 *If s_n is a non-constant hypergeometric term then $s_{n+1} - s_n$ is a hypergeometric term similar to s_n.*

Proof. Let $r(n) = s_{n+1}/s_n$. Then $s_{n+1} - s_n = (r(n) - 1)s_n$ is a nonzero rational multiple of s_n. $\square$

Proposition 5.6.2 *Let s_n and t_n be hypergeometric terms such that $s_n + t_n \neq 0$. Then $s_n + t_n$ is hypergeometric if and only if $s_n \sim t_n$.*

Proof. Let $a(n) = s_{n+1}/s_n$, $b(n) = t_{n+1}/t_n$, $c(n) = (s_{n+1} + t_{n+1})/(s_n + t_n)$, and $r(n) = s_n/t_n$. Then $a(n)$ and $b(n)$ are rational functions of n, and

$$c(n) = \frac{a(n)r(n) + b(n)}{r(n) + 1}, \tag{5.6.1}$$

so $c(n)$ is rational when $r(n)$ is.

Conversely, if $c(n) = a(n)$ then it follows from (5.6.1) that $a(n) = b(n)$, hence s_n and t_n are constant multiples of each other and $r(n)$ is constant. If $c(n) \neq a(n)$ then from (5.6.1)

$$r(n) = \frac{b(n) - c(n)}{c(n) - a(n)}.$$

In either case, $r(n)$ is rational when $c(n)$ is. $\square$

Theorem 5.6.1 *Let* $t_n^{(1)}, t_n^{(2)}, \ldots, t_n^{(k)}$ *be hypergeometric terms such that*

$$\sum_{i=1}^{k} t_n^{(i)} = 0. \tag{5.6.2}$$

Then $t_n^{(i)} \sim t_n^{(j)}$ *for some* i *and* j, $1 \leq i < j \leq k$.

Proof. We prove the assertion by induction on k.

If $k = 1$, then $t_n^{(1)} \neq 0$, since hypergeometric terms are nonzero by definition.

If $k > 1$, let $r_i(n) = t_{n+1}^{(i)}/t_n^{(i)}$, for $i = 1, 2, \ldots, k$. From (5.6.2) it follows that $\sum_{i=1}^{k} t_{n+1}^{(i)} = 0$, too, so

$$\sum_{i=1}^{k} r_i(n) t_n^{(i)} = 0. \tag{5.6.3}$$

Multiply (5.6.2) by $r_k(n)$ and subtract (5.6.3) to find

$$\sum_{i=1}^{k-1} (r_k(n) - r_i(n)) t_n^{(i)} = 0. \tag{5.6.4}$$

If $r_k(n) - r_i(n) = 0$ for some i, then $t_n^{(k)}/t_n^{(i)}$ is constant and hence $t_n^{(k)} \sim t_n^{(i)}$. Otherwise all terms on the left of (5.6.4) are hypergeometric, so by the induction hypothesis there are i and j, $1 \leq i < j \leq k - 1$, such that $(r_k(n) - r_i(n)) t_n^{(i)} \sim (r_k(n) - r_j(n)) t_n^{(j)}$. But then $t_n^{(i)} \sim t_n^{(j)}$ as well. $\square$

Proposition 5.6.3 *Every sum of a fixed number of hypergeometric terms can be written as a sum of pairwise dissimilar hypergeometric terms.*

Proof. Since the sum of two similar hypergeometric terms is either hypergeometric or zero, this can be achieved by grouping together similar terms. Each such group is a single hypergeometric term, by Proposition 5.6.2. $\square$

How do we decide if two hypergeometric terms are similar? This reduces to the question whether a given hypergeometric term is rational or not. In practice, this will be decided by an appropriate simplification routine for hypergeometric terms (such as our `FactorialSimplify`, for example). But since all computation with hypergeometric terms can be translated into corresponding operations with their rational function representations, we show how to decide rationality of a hypergeometric term given only its consecutive-term ratio.

Theorem 5.6.2 *Let t_n be a hypergeometric term and $r(n) = t_{n+1}/t_n$ its rational consecutive-term ratio. Let*

$$r(n) = \frac{A(n)}{B(n)} \frac{C(n+1)}{C(n)}$$

and

$$\frac{B(n)}{A(n)} = \frac{a(n)}{b(n)} \frac{c(n+1)}{c(n)} \tag{5.6.5}$$

be the canonical factorizations of $r(n)$ and of $B(n)/A(n)$, respectively, as described in Theorem 5.3.1. Then t_n is a rational function of n if and only if $A(n)$ is monic and $a(n) = b(n) = 1$.

Proof. If $a(n) = b(n) = 1$ then $r(n) = c(n)C(n+1)/(c(n+1)C(n))$, so $t_n = \alpha C(n)/c(n)$ where $\alpha \in K$ is some constant. Hence t_n is rational.

Conversely, assume that $t_n = p(n)/q(n)$ where p, q are relatively prime polynomials and q is monic. Then

$$\frac{q(n)}{q(n+1)} \frac{p(n+1)}{p(n)} = \frac{A(n)}{B(n)} \frac{C(n+1)}{C(n)}. \tag{5.6.6}$$

Obviously, $A(n)$ is monic. By Lemma 5.3.1, $p(n)$ divides $C(n)$. Write $C(n) = p(n)s(n)$, where $s(n)$ is a polynomial. Then by (5.6.6)

$$\frac{B(n)}{A(n)} = \frac{q(n+1)}{q(n)} \frac{s(n+1)}{s(n)}.$$

By Corollary 5.3.1, factorization (5.6.5) is unique. Therefore $c(n) = q(n)s(n)$ and $a(n) = b(n) = 1$. □

Now we are ready to answer the questions posed at the start of this section.

1. Suppose that $\sum_{k=0}^{n-1} t_k = a_n^{(1)} + a_n^{(2)} + \cdots + a_n^{(m)}$ where $a_n^{(i)}$ are hypergeometric terms. By Proposition 5.6.3 we can assume that these terms are pairwise dissimilar. Then $t_n = \Delta a_n^{(1)} + \Delta a_n^{(2)} + \cdots + \Delta a_n^{(m)}$. The nonzero terms on the right are pairwise dissimilar. By Theorem 5.6.1, there can be at most one nonzero term on the right. It follows that m above is at most 2, and if it is 2 then one of $a_n^{(1)}$, $a_n^{(2)}$ must be a constant. So the answer to question 1 is as follows.

Theorem 5.6.3 *If Gosper's algorithm does not succeed then the given sum cannot be expressed as a linear combination of a fixed number of hypergeometric terms (i.e., the sum is not expressible in closed form).*

Thus Gosper's algorithm in fact achieves more than it was designed for.

2. Similarly, the following algorithm will decide question 2 on page 94.

Extended Gosper's Algorithm

INPUT: Hypergeometric terms $t_n^{(1)}, t_n^{(2)}, \ldots, t_n^{(p)}$.

OUTPUT: Discrete functions $s_n^{(1)}, s_n^{(2)}, \ldots, s_n^{(q)}$ such that
$$\Delta \sum_{i=1}^{q} s_n^{(i)} = \sum_{j=1}^{p} t_n^{(j)};$$
  if at all possible, these functions will be hypergeometric terms.

Step 1. Write $\sum_{j=1}^{p} t_n^{(j)} = \sum_{j=1}^{q} u_n^{(j)}$ where $u_n^{(j)}$ are pairwise dissimilar.

Step 2. For $j = 1, 2, \ldots, q$ do:
  use Gosper's algorithm to find $s_n^{(j)}$ such that $\Delta s_n^{(j)} = u_n^{(j)}$;
  if Gosper's algorithm does not succeed then
    let $s_n^{(j)} = \sum_{k=0}^{n-1} u_k^{(j)}$.

Step 3. Return $\sum_{j=1}^{q} s_n^{(j)}$ and stop.   □

5.7   Exercises

1. [Gosp77] Evaluate the following sums.

   (a) $\sum_{n=0}^{m} n^k,$ for $k = 1, 2, 3, 4$

   (b) $\sum_{n=0}^{m} n^k 2^n,$ for $k = 1, 2, 3$

   (c) $\sum_{n=0}^{m} \frac{1}{n^2 + \sqrt{5}n - 1}$

   (d) $\sum_{n=0}^{m} \frac{n^4 4^n}{\binom{2n}{n}}$

   (e) $\sum_{n=0}^{m} \frac{(3n)!}{n!(n+1)!(n+2)!27^n}$

   (f) $\sum_{n=0}^{m} \frac{\binom{2n}{n}^2}{(n+1)4^{2n}}$

   (g) $\sum_{n=0}^{m} \frac{(4n-1)\binom{2n}{n}^2}{(2n-1)^2 4^{2n}}$

   (h) $\sum_{n=0}^{m} n \frac{(n-\frac{1}{2})!^2}{(n+1)!^2}$

2. [Gosp77] Find a closed form for the following sums containing parameters.

   (a) $\sum_{n=0}^{m} n^2 a^n$

   (b) $\sum_{n=0}^{m} (n - \frac{r}{2})\binom{r}{n}$

(c) $\sum_{n=1}^{m} \frac{(n-1)!^2}{(n-x)!(n+x)!}$

(d) $\sum_{n=0}^{m} \frac{n(n+a+b)a^n b^n}{(n+a)!(n+b)!}$

3. Express each of the following sums as a hypergeometric term plus a constant, or prove that they cannot be so expressed.

(a) $\sum_{n=1}^{m} \frac{1}{n^k}$, for $k = 1, 2, 3$

(b) $\sum_{n=1}^{m} \frac{6n+3}{4n^4+8n^3+8n^2+4n+3}$ [Abra71]

(c) $\sum_{n=1}^{m-1} \frac{n^2-2n-1}{n^2(n+1)^2} 2^n$

(d) $\sum_{n=1}^{m} \frac{n^2 4^n}{(n+1)(n+2)}$

(e) $\sum_{n=0}^{m-1} \frac{2^n}{n+1}$ [Man93]

(f) $\sum_{n=4}^{m} \frac{4(1-n)(n^2-2n-1)}{n^2(n+1)^2(n-2)^2(n-3)^2}$ [Man93]

(g) $\sum_{n=1}^{m} \frac{(n^4-14n^2-24n-9)2^n}{n^2(n+1)^2(n+2)^2(n+3)^2}$ [Man93]

(h) $\sum_{n=1}^{m} \frac{\prod_{j=1}^{n-1} j^3}{\prod_{j=1}^{n+1}(j^3+1)}$ [Gosp78]

(i) $\sum_{n=1}^{m} \frac{\prod_{j=1}^{n-1}(aj^3+bj^2+cj+d)}{\prod_{j=1}^{n}(aj^3+bj^2+cj+e)}$ (assuming $d \neq e$) [Gosp78]

(j) $\sum_{n=1}^{m} \frac{\prod_{j=1}^{n-1}(bj^2+cj+d)}{\prod_{j=1}^{n+1}(bj^2+cj+e)}$ (assuming $d \neq e$) [Gosp78]

4. Let $h_n = \binom{2n}{n} a^n$, where a is a parameter.

(a) Prove that h_n is not Gosper-summable over $\mathbf{Q}(a)$.

(b) Find all values of a for which h_n is Gosper-summable over $\mathbf{Q}$.

5. [Man93] Let K be a field of characteristic 0, $a \in K$, $a \neq 0$, and k a positive integer.

(a) Show that $f(n) = a^n/n^k$ is not Gosper-summable over K.

(b) Let $p(n)$ be a polynomial of degree $k-1$. Show that $f(n) = p(n)/\prod_{j=1}^{k}(n+a+j)$ is not Gosper-summable over K.

6. Let $f(n)$ be a sequence over some field. Define

$$f(n,s) = \sum_{n_s=0}^{n} \sum_{n_{s-1}=0}^{n_s} \cdots \sum_{n_1=0}^{n_2} f(n_1).$$

Show that $f(n,s) = \sum_{k=0}^{n} \binom{k+s-1}{k} f(n-k)$.

7. [PaSc94] Find a nonzero polynomial $p \in \mathbb{Q}(n)[k]$ of lowest degree such that $t_k = p(k)/k!$ is Gosper-summable.

8. Prove that unless the summand t_k is a rational function of k, equation (5.2.6) has at most one polynomial solution.

9. [LPS93] Show that in Step 3 of Gosper's algorithm Case 2b need not be considered when the summand t_k is a rational function of k.

10. [Petk94] Use Lemma 5.3.1 to derive Gosper's "miraculous" discovery (5.2.5) about the solution of (5.2.2).

11. [WZ90a] For each of the following hypergeometric terms $F(n,k)$ show that $F(n,k)$ is not Gosper-summable w.r.t. k. Then show that $F(n+1,k)-F(n,k)$ *is* Gosper-summable on k:

(a) $F(n,k) = \dfrac{\binom{n}{k}}{2^n}$,

(b) $F(n,k) = \dfrac{\binom{n}{k}^2}{\binom{2n}{n}}$,

(c) $F(n,k) = \dfrac{\binom{n}{k}n!}{k!(a-k)!(n+a)!}$,

(d) $F(n,k) = \dfrac{(-1)^k \binom{n+b}{n+k}\binom{n+c}{c+k}\binom{b+c}{b+k}}{\binom{n+b+c}{n,b,c}}$.

Solutions

1. (a) $\dfrac{m(1+m)}{2}$; $\dfrac{m(1+m)(1+2m)}{6}$; $\dfrac{m^2(1+m)^2}{4}$; $\dfrac{m(1+m)(1+2m)(-1+3m+3m^2)}{30}$

(b) $2 + 2^{m+1}(-1+m)$; $-6 + 2^{m+1}(3 - 2m + m^2)$;
$26 + 2^{m+1}(-13 + 9m - 3m^2 + m^3)$

(c) $\dfrac{m+1}{6}$ $\dfrac{m(m^2-7m+3)\sqrt{5}-(3m^3-7m^2+19m-6)}{2m^3\sqrt{5}+(m^4+5m^2-1)}$

(d) $-\dfrac{2}{231} + \dfrac{24^m(1+m)(3-22m+18m^2+112m^3+63m^4)}{693\binom{2m}{m}}$

(e) $-\dfrac{9}{2} + \dfrac{(200+261m+81m^2)(2+3m)!}{40\,27^m\,m!(1+m)!(2+m)!}$

(f) $\dfrac{(1+2m)^2\binom{2m}{m}^2}{4^{2m}(1+m)}$

(g) $-\dfrac{\binom{2m}{m}^2}{4^{2m}}$

(h) $4\pi - \dfrac{(1+2m)^2(4+3m)\left(-\frac{1}{2}+m\right)!^2}{(1+m)!^2}$

2. (a) $-\frac{a\,(1+a)}{(-1+a)^3} + \frac{a^{1+m}\left(1+a+2\,m-2\,a\,m+m^2-2\,a\,m^2+a^2\,m^2\right)}{(-1+a)^3}$

   (b) $\frac{\left(m-\frac{r}{2}\right)(-m+r)\binom{r}{m}}{-2\,m+r}$

   (c) $\frac{1}{(1-x)!\,(1+x)!} - \frac{1}{x^2\,(1-x)!\,(1+x)!} + \frac{m!^2}{x^2\,(m-x)!\,(m+x)!}$

   (d) $\frac{1}{(-1+a)!\,(-1+b)!} - \frac{a^{1+m}\,b^{1+m}}{(a+m)!\,(b+m)!}$

3. (a) Not Gosper-summable.

   (b) $\frac{m\,(2+m)}{3+4\,m+2\,m^2}$

   (c) $-2 + \frac{2^m}{m^2}$

   (d) $\frac{2}{3} + \frac{4^{m+1}\,(m-1)}{3\,(m+2)}$

   (e) Not Gosper-summable.

   (f) $-\frac{1}{16} + \frac{1}{(m-2)^2\,(m+1)^2}$

   (g) $-\frac{2}{9} + \frac{2^{m+1}}{(m+1)^2\,(m+3)^2}$

   (h) Not Gosper-summable.

   (i) $\frac{1}{d-e}\left(\frac{\prod_{j=1}^{m}(d+c\,j+b\,j^2+a\,j^3)}{\prod_{j=1}^{m}(e+c\,j+b\,j^2+a\,j^3)} - 1\right)$

   (j) $\frac{A+B}{(d-e)\,(-b^2+c^2-2\,b\,d-d^2-2\,b\,e+2\,d\,e-e^2)}$ where

$A = \frac{2\,b^2+2\,b\,c+3\,b\,d+c\,d+d^2-b\,e-c\,e-2\,d\,e+e^2}{b+c+e}$,

$B = \frac{\left(-2\,b^2-2\,b\,c-3\,b\,d-c\,d-d^2+b\,e+c\,e+2\,d\,e-e^2-4\,b^2\,m-2\,b\,c\,m-2\,b\,d\,m+2\,b\,e\,m-2\,b^2\,m^2\right)}{\left(\prod_{j=1}^{m+1}(e+c\,j+b\,j^2)\right)\left(\prod_{j=1}^{m}(d+c\,j+b\,j^2)\right)^{-1}}$.

4. (b) $a = 1/4$

6. Use induction on s.

7. $p(k) = k-1$

8. If (5.2.6) has two different polynomial solutions $x_1(n)$ and $x_2(n)$, then (5.1.3) has two different hypergeometric solutions $s_n^{(i)} = b(n-1)x_i(n)t_n/c(n)$, $i = 1, 2$. Their difference $s_n^{(1)} - s_n^{(2)}$ satisfies the homogeneous recurrence $z_{n+1} - z_n = 0$ and is therefore a (nonzero) constant. It follows that $s_n^{(1)} - s_n^{(2)}$ is hypergeometric. By Proposition 5.6.2, $s_n^{(1)}$ and $s_n^{(2)}$ are similar to a constant and are therefore rational. Hence t_k is rational, too.

9. If eq. (5.1.3) with rational t_n has a hypergeometric solution $z_n = s_n$, then s_n is rational and $s_n + C$ is a hypergeometric solution of (5.1.3) for any constant C. Then $x_n = c(n)(s_n + C)/(b(n-1)t_n)$ is a polynomial solution of (5.2.6)

for any constant C. Write $x_n = u(n) + Cv(n)$, where $u(n)$ and $v(n)$ are polynomials. It is easy to see that Case 1 does not apply. Since $v(n)$ is a nonzero solution of the homogeneous part of (5.2.6), its degree comes from Case 2b, because Cases 1 and 2a give $-\infty$ in the homogeneous case. Note that (5.2.6) has a polynomial solution whose degree is different from $\deg v$: If $\deg u \neq \deg v$ then this solution is $u(n)$, otherwise it is $u(n) - (\operatorname{lc} u/\operatorname{lc} v)v(n)$. Its degree must then come from Case 2a, so it is not necessary to examine Case 2b.

10. Let $y(n) = f(n)/g(n)$ where $f(n)$, $g(n)$ are relatively prime polynomials. Write $r(n)$ as in (5.2.3). Then, by (5.2.2),

$$\frac{a(n)}{b(n)} \frac{c(n+1)}{c(n)} = r(n) = \frac{y(n)+1}{y(n+1)} = \frac{f(n)+g(n)}{f(n+1)} \frac{g(n+1)}{g(n)}.$$

By Lemma 5.3.1, $g(n)$ divides $c(n)$, so $c(n)$ is a suitable denominator for $y(n)$. Write $y(n) = v(n)/c(n)$, and substitute this together with (5.2.3) into (5.2.2), to obtain $a(n)v(n+1) = (v(n) + c(n))b(n)$. This shows that $b(n)$ divides $v(n+1)$, hence $y(n) = b(n-1)x(n)/c(n)$ where $x(n)$ is a polynomial.

11. $F(n+1,k) - F(n,k) = G(n,k+1) - G(n,k)$ where $G(n,k) = R(n,k)F(n,k)$ and:

   (a) $R(n,k) = \frac{k}{2(k-n-1)}$,

   (b) $R(n,k) = \frac{(-3+2k-3n)k^2}{2(1+2n)(n-k+1)^2}$,

   (c) $R(n,k) = \frac{k^2}{(1+a+n)(k-n-1)}$,

   (d) $R(n,k) = -\frac{(b+k)(c+k)}{2(n+1-k)(n+1+b+c)}$.

Chapter 6

Zeilberger's Algorithm

6.1 Introduction

In the previous chapter we described Gosper's algorithm, which gives a definitive answer to the question of whether or not a given hypergeometric term can be *indefinitely* summed. That is, if $F(k)$ is such a term, we want to know if $F(k) = G(k+1) - G(k)$ where $G(k)$ is also such a term, and Gosper's algorithm fully answers that question.

In this chapter we study the algorithm that occupies a similarly central position in the study of *definite* sums, called Zeilberger's algorithm, or the method of *creative telescoping* [Zeil91, Zeil90b].

We are interested in a sum

$$f(n) = \sum_k F(n, k), \tag{6.1.1}$$

where $F(n, k)$ is a hypergeometric term in both arguments, i.e., $F(n+1, k)/F(n, k)$ and $F(n, k + 1)/F(n, k)$ are *both* rational functions of n and k. For the moment let's think of the range of the summation index k as being the set of all integers. Later we'll see that this assumption can be considerably relaxed.

What we want to find is a recurrence relation for the sum $f(n)$, and we'll do that by first finding a recurrence relation for the summand $F(n, k)$, just as in Sister Celine's algorithm of Chapter 4. So the method of creative telescoping is basically an alternative method of doing the same job that Sister Celine's algorithm does.

But it does that job a great deal faster.

Note first how different this question is from the one of Chapter 5. Certainly it is true that if $F(n, k) = G(n, k + 1) - G(n, k)$ for some nice G then we will easily be able to do our sum and find $f(n)$. But in this case we could do much more than merely find $f(n)$. We could actually express the sum as a function of a variable upper limit. But that is too much to expect in general. Many, many summands are not indefinitely summable, so Gosper's algorithm returns "No," but nevertheless the sum $f(n)$, where the index k runs over *all integers*, can be expressed in simple terms.

For instance, the binomial coefficient $\binom{n}{k}$, thought of for fixed n as a function of k, is not Gosper-summable. Nevertheless the unrestricted sum $\sum_k \binom{n}{k} = 2^n$ has a nice simple form, even though the indefinite sums $\sum_{k=0}^{K_0} \binom{n}{k}$ cannot be expressed as simple hypergeometric terms in K_0 (and n).

This situation is, of course, fully analogous to definite vs. indefinite integration. The function e^{-t^2} is not the derivative of any simple elementary function, so the indefinite integral $\int e^{-t^2} dt$ cannot be "done." Nonetheless the *definite* integral $\int_{-\infty}^{\infty} e^{-t^2} dt$ can be "done," and is equal to $\sqrt{\pi}$.

To return to our sum in (6.1.1), even though we cannot expect, in general, to find a term $G(n, k)$ such that $F(n, k) = G(n, k + 1) - G(n, k)$, we saw in Chapter 2, and we will see in more detail in Chapter 7, that often we get lucky and find a $G(n, k)$ for which

$$F(n + 1, k) - F(n, k) = G(n, k + 1) - G(n, k). \tag{6.1.2}$$

If that happens, then even though we can't do the indefinite sum of F, we can prove the definite summation identity $f(n) = \text{const}$.

In general, we cannot expect (6.1.2) to happen always either, but there is something that we *can* expect to happen, and we will prove that it does happen under very general circumstances. That is, we need to take a somewhat more general difference operator in n on the left side of (6.1.2).

Let N (resp. K) denote the forward shift operator in n (resp. k), i.e., $Ng(n, k) = g(n + 1, k)$, $Kg(n, k) = g(n, k + 1)$. In operator terms, then, (6.1.2) reads as $(N - 1)F = (K - 1)G$. We will show that we will "always" be able to find a difference operator of the form $p(n, N) = a_0(n) + a_1(n)N + a_2(n)N^2 + \cdots + a_J(n)N^J$ such that

$$p(n, N)F(n, k) = (K - 1)G(n, k),$$

in which the coefficients $\{a_i(n)\}_0^J$ are polynomials in n, and in which $G(n, k)/F(n, k)$ is a rational function of n, k, i.e., such that

$$\sum_{j=0}^{J} a_j(n)F(n + j, k) = G(n, k + 1) - G(n, k). \tag{6.1.3}$$

The mission of Zeilberger's algorithm, also known as the *method of creative tele-scoping*, is to produce the recurrence (6.1.3), given the summand function $F(n, k)$.

Suppose, for a moment, that we are trying to do the sum $f(n) = \sum_k F(n, k)$. Suppose that we execute Zeilberger's algorithm, and we find a recurrence of the form (6.1.3) for the summand function F, and a rational function $R(n, k)$ for which $G(n, k) = R(n, k)F(n, k)$. How does this help us to find the sum $f(n)$?

Since the coefficients on the left side of (6.1.3) are independent of k, we can sum (6.1.3) over all integer values of k and obtain

$$\sum_{j=0}^{J} a_j(n)f(n + j) = 0, \qquad (6.1.4)$$

assuming, say, that $G(n, k)$ has compact support in k for each n. Now there are several possible scenarios.

It might happen that $J = 1$, i.e., that equation (6.1.4) is a recurrence $a_0(n)f(n) + a_1(n)f(n+1) = 0$ of first order with polynomial coefficients. Well, then we're happy, because $f(n + 1)/f(n) = -a_0(n)/a_1(n)$ is a rational function of n, so our desired sum $f(n)$ is indeed a hypergeometric term, namely

$$f(n) = f(0) \prod_{j=0}^{n-1} (-a_0(j)/a_1(j)).$$

So in this case we have really done our sum.

It might happen that, even though $J > 1$ in the recurrence (6.1.3) that we find for our sum, we are lucky because the coefficients $\{a_i(n)\}_0^J$ are *constant*. Well, then we all know how to solve linear recurrence relations with constant coefficients, so once again we are assured that we will be able to find an explicit, simple formula for our sum $f(n)$.

It might be that neither of the above happens. Now you are looking at a recurrence formula in our unknown sum $f(n)$, with polynomial coefficients, and you have no idea how to solve it or if it can be solved, in any reasonable sense. Even in this difficult case, you are *certain* to obtain a complete answer to your question! The main algorithm of Chapter 8, Petkovšek's algorithm, deals definitively with exactly this situation. If your recurrence (6.1.4) has a solution $f(n)$ that is a linear combination of a fixed number of hypergeometric terms in n, then that algorithm will find the solution, otherwise it will return "No such solution exists."

Now we can go all the way back to the beginning. You are looking at a sum, $f(n) = \sum_k F(n, k)$, where F is a hypergeometric term, and you are wondering if there is a simple evaluation of that sum. If a "simple evaluation" means a formula for $f(n)$ that expresses it as a linear combination of a fixed number of hypergeometric terms, then the road to the answer is *completely algorithmic*, and is fully equipped with theorems that *guarantee* that either the algorithms will find a "simple evaluation" of your sum, or that your sum does not possess any such evaluation.

Hence the problem of evaluation of definite hypergeometric sums is, by means of Zeilberger's or Sister Celine's algorithm together with Petkovšek's algorithms, placed in the elite class that contains, for example, the problem of indefinite integration in the Liouvillian sense, and the question of indefinite hypergeometric summation, viz., the class of famous questions of classical mathematics that turn out to have completely algorithmic solutions.

Example 6.1.1. Let's try, as an example, problem 10424 from *The American Mathematical Monthly* (the methods of this book are great for a lot of Monthly problems!). It calls for the evaluation of the sum

$$f(n) = \sum_{0 \le k \le n/3} 2^k \frac{n}{n-k} \binom{n-k}{2k}.$$

If we simply give the summand $F(n, k) = 2^k \frac{n}{n-k} \binom{n-k}{2k}$ to the *creative telescoping* algorithm, as implemented by program ct in the EKHAD package of Maple programs that accompanies this book, it very quickly responds by telling us that the summand F satisfies the third order recurrence

$$(N^2 + 1)(N - 2)F(n, k) = G(n, k+1) - G(n, k), \tag{6.1.5}$$

where

$$G(n, k) = -\frac{2^k n}{n - 3k + 3} \binom{n-k}{2k-2}.$$

If we sum the recurrence over $0 \le k \le n - 1$, then for $n \ge 2$ the right side telescopes to 0 (check this carefully!), and we find that the unknown sum satisfies $(N^2 + 1)(N - 2)f(n) = 0$. But that is a recurrence with *constant* coefficients, and, furthermore, it is one whose general solution is clearly $f(n) = c_1 2^n + c_2 i^n + c_3 (-i)^n$. If we match $f(1), f(2), f(3)$ to this formula we obtain the complete evaluation

$$f(n) = 2^{n-1} + \frac{1}{2}(i^n + (-i)^n) = 2^{n-1} + \cos \frac{n\pi}{2}$$

for $n \ge 2$, and the case $n = 1$ can be checked separately.

This example was typical in some respects and atypical in others. It was typical in that the algorithm returned a recurrence relation of the form (6.1.3) for the summand. It was atypical in that the recurrence has constant coefficients and order 3. □

6.2   Existence of the telescoped recurrence

In this and the next section we will study the existence and the implementation of
the algorithm.

The *existence* of a recurrence of the form (6.1.3) for the summand $F(n,k)$ is
assured, under the same hypotheses as those of Theorem 4.4.1 (the Fundamental Theorem), namely that $F(n,k)$ should be a proper hypergeometric term (see
page 64).

The proof of the existence will follow at once from Theorem 4.4.1, which assures
the existence of the two-variable recurrence (4.4.2).

The implementation of the creative telescoping algorithm is very different from
that of Sister Celine's algorithm. In principle, one could first find the two-variable
recurrence (4.4.2) and then proceed as in the proof of Theorem 6.2.1 below to
convert it into a recurrence in the telescoped form (6.1.3). But Zeilberger found a
much more efficient way to implement it, a procedure that uses a variant of Gosper's
algorithm, as we will see.

Theorem 6.2.1 *Let $F(n,k)$ be a proper hypergeometric term. Then F satisfies
a nontrivial recurrence of the form (6.1.3), in which $G(n,k)/F(n,k)$ is a rational
function of n and k.*

Proof.   The proof, following Wilf and Zeilberger [WZ92a], begins with the two-
variable recurrence (4.4.2) which we repeat here, in the form

$$\sum_{i=0}^{I}\sum_{j=0}^{J} a_{i,j}(n)F(n+j,k+i) = 0. \qquad (6.2.1)$$

We know that such a recurrence exists nontrivially. Introduce the shift operators
K, N, defined, as usual, by $Ku(k) = u(k+1)$ and $Nv(n) = v(n+1)$. Then (6.2.1)
can be written in operator form as $P(N,n,K)F(n,k) = 0$. Suppose we take the
polynomial $P(u,v,w)$ and expand it in a power series in w, about the point $w = 1$,
to get

$$P(u,v,w) = P(u,v,1) + (1-w)Q(u,v,w),$$

where Q is a polynomial. Then (6.2.1) implies that

$$0 = P(N,n,K)F(n,k) = (P(N,n,1) + (1-K)Q(N,n,K))F(n,k),$$

i.e., that

$$P(N,n,1)F(n,k) = (K-1)Q(N,n,K)F(n,k). \qquad (6.2.2)$$

On the left side of this latter recurrence, only n varies. On the right side, if we put $G(n,k) = Q(N,n,K)F(n,k)$, then the right side is simply $G(n,k+1)-G(n,k)$, and G is itself a rational function multiple of F, since any number of shift operators, when applied to a hypergeometric term, only multiply it by a rational function.

We claim finally that the recurrence (6.2.2) is nontrivial. The following proof is due to Graham, Knuth and Patashnik [GKP89].

We know by the Fundamental Theorem that there are operators $P(N,n,K)$, which are nontrivial, which depend only on N,n,K and which annihilate $F(n,k)$. Among these, let $P = P(N,n,K)$ be one that has the least degree in K. Divide P by $K-1$ to get

$$P(N,n,K) = P(N,n,1) - (K-1)Q(N,n,K),$$

which defines the operator Q.

Suppose $P(N,n,1) = 0$. Then $(K-1)G(n,k) = 0$, i.e., G is independent of k. Hence G is a hypergeometric term in the single variable n, i.e., G satisfies a recurrence of order 1 with polynomial-in-n coefficients. Thus there is a first-order operator $H(N,n)$ such that $H(N,n)G(n,k) = 0$.

If $Q = 0$ then $P(N,n,K) = P(N,n,1)$ is a nonzero k-free operator that is independent of K and k and annihilates $F(n,k)$. If $Q \neq 0$, then $H(N,n)Q(N,n,K)$ is a nonzero k-free operator annihilating $F(n,k)$.

In either case we have found a nonzero k-free operator that annihilates $F(n,k)$ and whose degree in K is smaller than that of $P(N,n,K)$, which is a contradiction, since P was assumed to have minimum degree in K among such operators.   □

Hence a recurrence in telescoped form always exists. It can be found by rearranging the terms in the two-variable Sister Celine form of the recurrence, but we will now see that there is a much faster way to get the job done.

6.3   How the algorithm works

The creative telescoping algorithm is for the fast discovery of the recurrence for a proper hypergeometric term, in the telescoped form (6.1.3). The algorithmic implementation makes strong use of the existence, but not of the method of proof used in the existence theorem.

More precisely, what we do is this. We now *know* that a recurrence (6.1.3) exists. On the left side of the recurrence there are unknown coefficients $a_0, \ldots, a_J$; on the right side there is an unknown function G; and the order J of the recurrence is unknown, except that bounds for it were established in the Fundamental Theorem (Theorem 4.4.1 on page 65).

We begin by fixing the assumed order J of the recurrence. We will then look for a recurrence of that order, and if none exists, we'll look for one of the next higher order.

For that fixed J, let's denote the left side of (6.1.3) by t_k, so that

$$t_k = a_0 F(n, k) + a_1 F(n + 1, k) + \cdots + a_J F(n + J, k). \qquad (6.3.1)$$

Then we have for the term ratio

$$\frac{t_{k+1}}{t_k} = \frac{\sum_{j=0}^{J} a_j F(n + j, k + 1)/F(n, k + 1)}{\sum_{j=0}^{J} a_j F(n + j, k)/F(n, k)} \frac{F(n, k + 1)}{F(n, k)}. \qquad (6.3.2)$$

The second member on the right is a rational function of n, k, say

$$\frac{F(n, k + 1)}{F(n, k)} = \frac{r_1(n, k)}{r_2(n, k)},$$

where the r's are polynomials, and also

$$\frac{F(n, k)}{F(n - 1, k)} = \frac{s_1(n, k)}{s_2(n, k)},$$

say, where the s's are polynomials. Then

$$\frac{F(n + j, k)}{F(n, k)} = \prod_{i=0}^{j-1} \frac{F(n + j - i, k)}{F(n + j - i - 1, k)} = \prod_{i=0}^{j-1} \frac{s_1(n + j - i, k)}{s_2(n + j - i, k)}. \qquad (6.3.3)$$

It follows that

$$\frac{t_{k+1}}{t_k} = \frac{\sum_{j=0}^{J} a_j \left\{ \prod_{i=0}^{j-1} \frac{s_1(n+j-i,k+1)}{s_2(n+j-i,k+1)} \right\} \frac{r_1(n, k)}{r_2(n, k)}}{\sum_{j=0}^{J} a_j \left\{ \prod_{i=0}^{j-1} \frac{s_1(n+j-i,k)}{s_2(n+j-i,k)} \right\}}$$

$$= \frac{\sum_{j=0}^{J} a_j \left\{ \prod_{i=0}^{j-1} s_1(n + j - i, k + 1) \prod_{r=j+1}^{J} s_2(n + r, k + 1) \right\}}{\sum_{j=0}^{J} a_j \left\{ \prod_{i=0}^{j-1} s_1(n + j - i, k) \prod_{r=j+1}^{J} s_2(n + r, k) \right\}} \qquad (6.3.4)$$

$$\times \frac{r_1(n, k)}{r_2(n, k)} \frac{\prod_{r=1}^{J} s_2(n + r, k)}{\prod_{r=1}^{J} s_2(n + r, k + 1)}.$$

Thus we have

$$\frac{t_{k+1}}{t_k} = \frac{p_0(k + 1)}{p_0(k)} \frac{r(k)}{s(k)}, \qquad (6.3.5)$$

where

$$p_0(k) = \sum_{j=0}^{J} a_j \left\{ \prod_{i=0}^{j-1} s_1(n + j - i, k) \prod_{r=j+1}^{J} s_2(n + r, k) \right\}, \qquad (6.3.6)$$

and

$$r(k) = r_1(n, k) \prod_{r=1}^{J} s_2(n + r, k), \qquad (6.3.7)$$

$$s(k) = r_2(n, k) \prod_{r=1}^{J} s_2(n + r, k + 1). \qquad (6.3.8)$$

Note that *the assumed coefficients a_j do not appear in $r(k)$ or in $s(k)$*, but only in $p_0(k)$.

Next, by Theorem 5.3.1, we can write $r(k)/s(k)$ in the canonical form

$$\frac{r(k)}{s(k)} = \frac{p_1(k + 1)}{p_1(k)} \frac{p_2(k)}{p_3(k)}, \qquad (6.3.9)$$

in which the numerator and denominator on the right are coprime, and

$$\gcd(p_2(k), p_3(k + j)) = 1 \quad (j = 0, 1, 2, \ldots).$$

Hence if we put $p(k) = p_0(k)p_1(k)$ then from eqs. (6.3.5) and (6.3.9), we obtain

$$\frac{t_{k+1}}{t_k} = \frac{p(k + 1)}{p(k)} \frac{p_2(k)}{p_3(k)}. \qquad (6.3.10)$$

This is now a standard setup for Gosper's algorithm (compare it with the discussion on page 78), and we see that t_k will be an indefinitely summable hypergeometric term if and only if the recurrence (compare eq. (5.2.6))

$$p_2(k)b(k + 1) - p_3(k - 1)b(k) = p(k) \qquad (6.3.11)$$

has a polynomial solution $b(k)$.

The remarkable feature of this equation (6.3.11) is that *the coefficients $p_2(k)$ and $p_3(k)$ are independent of the unknowns $\{a_j\}_{j=0}^{J}$, and the right side $p(k)$ depends on them linearly.* Now watch what happens as a result. We look for a polynomial solution to (6.3.11) by first, as in Gosper's algorithm, finding an upper bound on the degree, say Δ, of such a solution. Next we assume $b(k)$ as a general polynomial of that degree, say

$$b(k) = \sum_{l=0}^{\Delta} \beta_l k^l,$$

with all of its coefficients to be determined. We substitute this expression for $b(k)$ in (6.3.11), and we find a system of simultaneous *linear* equations in the $\Delta + J + 2$ unknowns

$$a_0, a_1, \ldots, a_J, \beta_0, \ldots, \beta_\Delta.$$

The linearity of this system is directly traceable to the italicized remark above.

We then solve the system, if possible, for the a_j's and the β_l's. If no solution exists, then there is no recurrence of telescoped form (6.1.3) and of the assumed order J. In such a case we would next seek such a recurrence of order $J+1$. If on the other hand a polynomial solution $b(k)$ of equation (6.3.11) does exist, then we will have found all of the a_j's of our assumed recurrence (6.1.3), and, by eq. (5.2.5) we will also have found the $G(n,k)$ on the right hand side, as

$$G(n,k) = \frac{p_3(k-1)}{p(k)} b(k) t_k. \qquad (6.3.12)$$

See Koornwinder [Koor93] for further discussion and a q-analogue.

6.4   Examples

Example 6.4.1. Now let's do by hand an example of the implementation of the creative telescoping algorithm, as described in the previous section. We take the summand $F(n,k) = \binom{n}{k}^2$ and try to find a recurrence of order $J = 1$.

For the term ratio, we have from (6.3.1) that

$$
\frac{t_{k+1}}{t_k} = \frac{a_0 \binom{n}{k+1}^2 + a_1 \binom{n+1}{k+1}^2}{a_0 \binom{n}{k}^2 + a_1 \binom{n+1}{k}^2} = \frac{a_0 \frac{(n-k)^2}{(k+1)^2} + a_1 \frac{(n+1)^2}{(k+1)^2}}{a_0 + a_1 \frac{(n+1)^2}{(n+1-k)^2}}
$$
$$
= \left\{ \frac{a_0(n-k)^2 + a_1(n+1)^2}{a_0(n+1-k)^2 + a_1(n+1)^2} \right\} \left\{ \frac{(n+1-k)^2}{(k+1)^2} \right\}, \qquad (6.4.1)
$$

which is of the form (6.3.5) with

$$p_0(k) = a_0(n-k+1)^2 + a_1(n+1)^2, \quad r(k) = (n+1-k)^2, \quad s(k) = (k+1)^2.$$

Now the canonical form (6.3.9) is simply

$$\frac{r(k)}{s(k)} = \frac{1}{1} \frac{(n+1-k)^2}{(k+1)^2},$$

i.e., we have

$$p_1(k) = 1, \quad p_2(k) = (n+1-k)^2, \quad p_3(k) = (k+1)^2.$$

Hence we put $p(k) = p_0(k)p_1(k) = a_0(n-k+1)^2 + a_1(n+1)^2$, and (6.3.10) takes the form (in this case identical with (6.4.1) above)

$$\frac{t_{k+1}}{t_k} = \frac{a_0(n-k)^2 + a_1(n+1)^2}{a_0(n-k+1)^2 + a_1(n+1)^2} \frac{(n+1-k)^2}{(k+1)^2}.$$

Now we must solve the recurrence (6.3.11), which in this case looks like

$$(n - k + 1)^2 b(k + 1) - k^2 b(k) = a_0(n - k + 1)^2 + a_1(n + 1)^2. \qquad (6.4.2)$$

More precisely, we want to know if there exist $a_0(n), a_1(n)$ such that this recurrence has a polynomial solution $b(k)$.

At this point in Gosper's algorithm, the next thing to do is to find an upper bound for the degree of a polynomial solution, if one exists. We observe first that we are here in Case 2 of the degree-bounding process that was described on page 86, and that the degree bound is 1. Therefore if there is any polynomial solution $b(k)$ at all, there is one of first degree.

Hence we assume that $b(k) = \alpha + \beta k$, where α, β (along with a_0, a_1) are to be determined, we substitute into the recurrence (6.4.2), and we match the coefficients of like powers of k on both sides. The result is that the choices

$$\alpha = -3(n + 1), \quad \beta = 2, \quad a_0 = -2(2n + 1), \quad a_1 = n + 1$$

do indeed satisfy (6.4.2). Thus $F(n, k) = \binom{n}{k}^2$ satisfies the telescoped recurrence

$$-2(2n + 1)F(n, k) + (n + 1)F(n + 1, k) = G(n, k + 1) - G(n, k) \qquad (6.4.3)$$

where, by (6.3.12),

$$G(n, k) = \frac{(2k - 3n - 3)n!^2}{(k - 1)!^2(n - k + 1)!^2}.$$

Now that the algorithm has returned the recurrence in telescoped form, it is quite easy to solve the recurrence for the sum $f(n) = \sum_k \binom{n}{k}^2$ and thereby to evaluate it. Indeed, if we sum the recurrence (6.4.3) over all integers k, the right side collapses to 0, and we find that

$$-2(2n + 1)f(n) + (n + 1)f(n + 1) = 0.$$

This, together with $f(0) = 1$ quickly yields the desired evaluation $f(n) = \binom{2n}{n}$.

□

That was the only example that we'll work out by hand. Here are a number of machine-generated examples that show more of what the algorithm can do.

Example 6.4.2. First we try to evaluate the sum

$$f(n) = \sum_k (-1)^k \frac{\binom{n}{k}}{\binom{x+k}{k}}.$$

If $F(n, k)$ denotes the summand, then the creative telescoping algorithm finds the recurrence

$$(-n - x)F(n, k) = G(n, k + 1) - G(n, k),$$

where $G = RF$ and $R(n, k) = (x + k)$. Summing over k, we find that our unknown sum $f(n)$ satisfies

$$(-x - n)f(n) = -G(n, 0) = -x,$$

and so $f(n) = x/(x+n)$. This was a case in which we needed Zeilberger's algorithm with ORDER:=0, i.e., Gosper's algorithm. The $F(n, k)$ here is *indefinitely* summable. That means that not only is there a closed form expression for $f(n)$, but also there is one for

$$f(n, K) = \sum_{k=0}^{K} (-1)^k \frac{\binom{n}{k}}{\binom{x+k}{k}},$$

for all K, namely

$$(-x - n)f(n, K) = G(n, K + 1) - G(n, 0) = (x + K + 1)(-1)^{K+1} \frac{\binom{n}{K+1}}{\binom{x+K+1}{K+1}} - x.$$

This example highlights the fact that when using the creative telescoping program one should always start with ORDER:=0, and increase it by 1 as necessary, or else use the routine zeil in the EKHAD package, which does this automatically. If we begin, instead, with ORDER:=1, we might miss a sum which has closed form for every value of its upper limit, instead of only for one value. $\square$

Example 6.4.3. Next, let's evaluate the sum

$$f(n) = \sum_k (-1)^k \binom{n + 1}{k} \binom{2n - 2k + 1}{n}.$$

If $F(n, k)$ is the summand, then algorithm ct quickly finds the recurrence

$$(N - 1)F(n, k) = G(n, k + 1) - G(n, k), \tag{6.4.4}$$

where $G = RF$ and

$$R(n, k) = -2 \frac{(3n + 6 - 2k)(n + 1 - k)(2n + 3 - 2k)k}{(n + 1)(n + 2 - k)(n + 2 - 2k)}.$$

(Remember: you don't have to take our word for it: substitute F, G, and see for yourself that (6.4.4) is true!) If we sum the recurrence over k, we find that our unknown sum satisfies $f(n + 1) = f(n)$, i.e., $f(n)$ is constant. Since $f(0) = 1$ we have shown that $f(n) = 1$ for all $n \geq 0$. Note that in this case we have actually found a WZ-style proof, so we could next look for companion and dual identities etc. □

Example 6.4.4. Now let's do the famous sum of Dixon,

$$f(n) = \sum_k (-1)^k \binom{2n}{k}^3.$$

Here the algorithm returns the recurrence

$$(-3(3n+2)(3n+1) - (n+1)^2 N)F(n,k) = G(n,k+1) - G(n,k),$$

where $G = RF$, and $R(n,k)$ is a pretty complicated rational function that we will not reproduce here. If we sum the recurrence over k, we find that the sum satisfies

$$(-3(3n+2)(3n+1) - (n+1)^2 N)f(n) = 0,$$

i.e.,

$$f(n+1) = -\frac{3(3n+2)(3n+1)}{(n+1)^2}f(n),$$

which, with $f(0) = 1$, easily yields the evaluation $f(n) = (-1)^n (3n)!/n!^3$. □

We could go on forever this way, proving one sum evaluation after another. Instead we'll defer a few of the examples to the next two sections, where you will see the programmed implementations of the algorithm at work.

A continuous analogue, that computes the linear recurrence satisfied by

$$a_n := \int F(n,y)dy,$$

where $F(n,y)$ is proper in the sense that both

$$\frac{F(n+1,y)}{F(n,y)} \quad \text{and} \quad \frac{D_y F(n,y)}{F(n,y)}$$

are rational functions of (n,y), and the differential equation satisfied by

$$f(x) := \int F(x,y)dy,$$

where F is such that both $D_x F/F$ and $D_y F/F$ are rational functions of (x,y), can be found in [AlZe90].

Procedures AZd, AZc of EKHAD are Maple implementations of these algorithms. The procedures AZpapd, AZpapc are verbose versions. In EKHAD, type help(AZpapd) etc. for details.

6.5   Use of the programs

In this section we will use two particular programs that implement the creative telescoping algorithm. The first one is Zeilberger's Maple program. The second is a Mathematica program, written by Peter Paule and Markus Schorn, of RISC-Linz, which is available from the WorldWideWeb, as described in Appendix A.

The Maple program

Zeilberger's program, in Maple, is program `ct` in the package `EKHAD` that accompanies this book (see Appendix A).

Suppose that $F(n, k)$ is a given summand for which you are interested in finding a recurrence for $f(n) = \sum_k F(n, k)$. Then make a call for `ct(SUMMAND,ORDER,k,n,N)`, where SUMMAND is your summand function, ORDER is the order of the recurrence that you are looking for, and `k,n` are respectively the summation and the running indices.

If no recurrence of the desired order exists, the output will say so. Otherwise the program will output

1. the desired recurrence for the summand $F(n, k)$, in the telescoped form (6.1.3),

2. the proof that the output recurrence is correct, namely the function $G(n, k)$ from which one can check the truth of (6.1.3),

3. the desired recurrence for the sum $f(n)$.

For instance, the program returns the recurrence for

$$f(n) = \sum_k \binom{n}{2k}\binom{2k}{k}4^{-k},$$

in the form `(2n+1+(-n-1)N)SUM(n)=0`. Here N is the forward shift operator on n, so in customary mathematical notation, the recurrence that the program found is

$$(2n + 1)f(n) - (n + 1)f(n + 1) = 0.$$

Before we list the rest of the output, we should note that we already have a complete evaluation of the sum $f(n)$ in closed form. Indeed, since

$$f(n + 1) = \frac{2n + 1}{n + 1}f(n) \qquad (n \geq 0;\ f(0) = 1),$$

one quickly finds that

$$f(n) = \sum_k \binom{n}{2k}\binom{2k}{k} 4^{-k} = 2^{-(n-1)}\binom{2n-1}{n-1} \quad (n \ge 1),$$

which is quite an effortless way to evaluate a binomial coefficient sum, we hope you'll agree.

To continue with the program output, we see also the function $G(n, k)$, on the right side of (6.1.3), which is

$$G(n,k) = \frac{n!}{(n+1-2k)!\, 4^{k-1}(k-1)!\,^2}.$$

Finally, there appears the full telescoping recurrence (6.1.3) in the form

$$(2n + 1 + (-n - 1)N)F(n, k) = G(n, k + 1) - G(n, k). \qquad (6.5.1)$$

It should be noted that this recurrence is *self-certifying*, which is to say that it proves itself. One does not need to take the computer's word for the truth of (6.5.1). To prove it we need only divide through by $F(n, k)$, cancel out the factorials, and check the correctness of the resulting polynomial identity.

The recurrences that are output by the creative telescoping algorithm (or for that matter, by Sister Celine's algorithm) are always self-certifying, in this sense. Humans may be hard put to *find* them, but we can *check* them easily.

The Mathematica program

The program of Peter Paule and Markus Schorn carries out the creative telescoping algorithm, in Mathematica. Their package zb_alg.m (get it through our Web page; see page 201) is first loaded by the Mathematica command <<zb_alg.m. Then several new commands become available, one of which is

Zb[summand,sumvar,runningvar,order]

and that is the one that finds the telescoped recurrence. Here summand is the summand $f(n, k)$, sumvar is the dummy index of summation, say k, runningvar is the variable in terms of which the recurrence for the sum will be found, say n, and order is the order of the recurrence that is being sought. The program is in many situations remarkably fast. It uses its own algorithms for finding null spaces of symbolic matrices.

Example 6.5.1. Let's try the program on the mystery sum (this is the Reed–Dawson sum that we evaluated on page 63)

$$f(n) = \sum_k \binom{n}{k}\binom{2k}{k}\left(-\frac{1}{2}\right)^k.$$

Here the summand is $F(n,k) = \binom{2k}{k}\binom{n}{k}(-\frac{1}{2})^k$, hence we call

    Zb[Binomial[2k,k] Binomial[n,k] (-1/2)^k,k,n,1]

The program replies: "**Try higher order**," i.e., no recurrence of first order was found. So we try again, this time calling

    Zb[Binomial[2k,k] Binomial[n,k] (-1/2)^k,k,n,2]

and we are answered with the recurrence

    (1 + n) SUM[n] + (-2 - n) SUM[2 + n] == 0.

Now, usually when a recurrence order higher than the first is found, it means that we will not be able to determine analytically whether or not a closed form solution exists, except by calling Petkovšek's algorithm Hyper, which is described in Chapter 8. In this case, however, the second order recurrence

$$f(n+2) = \frac{n+1}{n+2}f(n),$$

together with the initial values $f(0) = 1$, $f(1) = 0$, is easily solved by inspection, yielding that $f(2n+1) = 0$ and $f(2n) = 4^{-n}\binom{2n}{n}$, for integer n. □

Example 6.5.2. Let's find a closed form expression for the sum

$$f(n) = \sum_k \binom{n+k}{2k}\binom{2k}{k}\frac{(-1)^k}{k+1}.$$

We begin by calling the Paule–Schorn program with

    Zb[(n+k)! (-1)^k/(k! (k+1)! (n-k)!),k,n,1].

This time it replies with the recurrence

    (n (n + 1) SUM(n) - (n + 2) (n + 1) SUM(n+1) == 0.

Clearly $f(0) = 1$ here, and the recurrence shows that $f(n) = 0$ for all $n \geq 1$. Wasn't that painless? □

Example 6.5.3. In this example we will see how the algorithm can prove a difficult identity in linear algebra, a fact which opens the door to many future applications in that subject. This example is taken from Petkovšek and Wilf [PeW96]. The identity in question is due to Mills, Robbins and Rumsey [MRR87], and it gives the evaluation of the $n \times n$ determinant

$$\det \left\{ \binom{i+j-x}{2i-j} \right\}_{i,j=0,\ldots,n-1} \tag{6.5.2}$$

which occurs in the theory of plane partitions. Instead of giving the exact evaluation here, let us make the following remark. The determinant in (6.5.2) is clearly a polynomial in x. What is not clear, but is true, is that all of its roots are either integers or half-integers, so it has a complete factorization into factors of the form $x - a$, where $2a \in \mathbb{Z}$. For example, when $n = 4$, the evaluation is

$$\frac{(-5+x)\,(-4+x)\,(-3+x)\,(-2+x)\,(-17+2\,x)\,(-15+2\,x)}{180}.$$

How can the computer methods that we have developed for the proof of hypergeometric identities be adapted to prove a determinant evaluation?

Well, let M_n be the matrix in (6.5.2). Suppose we could exhibit a triangular matrix E_n, with 1's on the diagonal, for which $M_n E_n$ is triangular. Then the determinant of M_n would be the product of the diagonal entries of that triangular matrix. By computer experimentation, George Andrews [Andr98] discovered a matrix E_n that seemed to work (and he proved, by non-computer methods, that it *does* work), namely the matrix $E_n = (e_{i,j}(x))_{i,j=0}^{n-1}$, where $e_{i,j} = 0$ if $i > j$, and

$$e_{i,j}(x) = \frac{1}{(-4)^{j-i}} \frac{(2j-i-1)!\,(2x+3j+1)!\,(x+i)!\,(x+i+j-\frac{1}{2})!}{(j-i)!\,(i-1)!\,(2x+2j+i+1)!\,(x+j)!\,(x+2j-\frac{1}{2})!}, \tag{6.5.3}$$

otherwise. But how can we *prove* that this matrix E_n really triangularizes M_n?

Since we know the exact forms of M_n and E_n, we can write out as explicit sums the matrix entries of the allegedly triangular matrix $M_n E_n$. Then we would have to show that the above-the-diagonal sums vanish, and that the sums for the diagonal entries are equal to certain polynomials in x that we won't write out here, but from which the theorem would follow.

At that moment we would be facing a standard problem, or rather two of them, in the theory of computerized proofs of hypergeometric identities. To prove the determinant evaluation we would have to prove those two identities. Zeilberger's algorithm produces such a proof, though not without an unpleasantly large certificate (it contains a polynomial with about 850 monomials in it!), and the problem is done. We refer the reader to [PeW96] for the details. □

The method of creative telescoping has a q-analogue. That is, given a summand $F(n,k)$ for which both $F(n+1,k)/F(n,k)$ and $F(n,k+1)/F(n,k)$ are rational functions of the two variables q^n and q^k, the algorithm will find a telescoped recurrence

$$\Omega(N,n)F(n,k) = G(n,k+1) - G(n,k), \tag{6.5.4}$$

in which $G(n,k) = R(n,k)F(n,k)$ and R is a rational function of q^n and q^k that will also be found by the program. This program is the routine `qzeil` in the package qEKHAD that accompanies this book (see Appendix A).

In connection with q identities, Peter Paule [Paul94] has made the following beautiful and effective observation. Recall that we have emphasized the importance of writing sums with unrestricted summation indices whenever possible, e.g., of writing $\sum_k \binom{n}{k}$ instead of $\sum_{k=0}^{n} \binom{n}{k}$, even though they both represent the same sum. The former notation emphasizes that the summation runs over all integers k, and this often simplifies subsequent manipulations that we may wish to carry out.

But consider the fact that *every* function $f(k)$ can be written as an even part plus an odd part,

$$f(k) = \frac{f(k) + f(-k)}{2} + \frac{f(k) - f(-k)}{2},$$

and consider also the fact that the unrestricted sum over the odd part obviously vanishes, i.e., $\sum_k (f(k) - f(-k))/2 = 0$. Consequently, for every summand f, one has

$$F(n) = \sum_k f(n,k) = \sum_k \frac{f(n,k) + f(n,-k)}{2}.$$

What could be the advantage of using the symmetrized summand instead of the original one? Just this: The *order* of the recurrence relation that is obtained for the symmetrized summand $F(n,k) = (f(n,k) + f(n,-k))/2$ might be dramatically lower than that for the original summand! Paule found, for instance, that one form of a famous identity of Rogers–Ramanujan, namely

$$\sum_k \frac{q^{k^2}}{(q;q)_k(q;q)_{n-k}} = \sum_k \frac{(-1)^k q^{(5k^2-k)/2}}{(q;q)_{n-k}(q;q)_{n+k}}, \tag{6.5.5}$$

which is certifiable by a recurrence of order 5, can in fact be certified by a recurrence of order 2 if it is first written in the symmetrical form

$$\sum_k \frac{2q^{k^2}}{(q;q)_k(q;q)_{n-k}} = \sum_k \frac{(-1)^k(1+q^k)q^{(5k^2-k)/2}}{(q;q)_{n-k}(q;q)_{n+k}}. \tag{6.5.6}$$

Indeed, here is the proof of (6.5.6): We claim that both sides of (6.5.6) are annihilated by the operator

$$A(N) := (1 - q^n) - (1 + q - q^n + q^{2n-1})N^{-1} + qN^{-2}.$$

The complete, human-verifiable proof of this fact simply exhibits the certificates

$$R_L(n,k) = -q^{2n-1}(1 - q^{n-k}) \quad \text{and} \quad R_R(n,k) = \frac{q^{2n+3k}}{1 + q^k}(1 - q^{n-k}).$$

We humans can then easily check that

$$A(N)F_L = G_L(n,k) - G_L(n,k-1) \quad \text{and} \quad A(N)F_R = G_R(n,k) - G_R(n,k-1),$$
$$(6.5.7)$$

where F_L, F_R are the summands on the left and right sides, respectively, of (6.5.6), $G_L = R_L F_L$, and $G_R = R_R F_R$. The proof of the claimed identity (6.5.6) now follows by summation of (6.5.7) over all integers k, and verification of the cases $n = 0, 1$. □

Aside from q-sums, there are examples of ordinary sums where this same method[1] has worked. Petkovšek has found that for the sums $\sum_k f(n,k,t)$, where

$$f(n,k,t) = \frac{1}{tk+1}\binom{tk+1}{k}\binom{tn-tk}{n-k}$$

one should symmetrize about $k = n/2$ by using the summand $(f(n,k,t) + f(n,n-k,t))/2$ instead. If one does that, then the orders of the recurrences obtained by the method of creative telescoping for the original summand and the symmetrized summand are as follows: for $t = 3$, 2 and 1; for $t = 4$, 4 and 2; and for $t = 5$, 6 and 3. Thus Paule's symmetrization method should be considered in cases where the recurrences are large and there is a natural point about which to symmetrize the summand.

6.6 Exercises

1. Use creative telescoping to evaluate each of the following binomial coefficient sums, in explicit closed form. In each case find a recurrence that is satisfied by the summand, then sum the recurrence over the range of the given summation to find a recurrence that is satisfied by the sum. Then solve that recurrence for the sum, either by inspection, or by being very clever, or, *in extremis*, by using algorithm Hyper of Chapter 8, page 156.

---

[1] Definition: A *method* is a trick that has worked at least twice.

(a) $\sum_k (-1)^k \binom{n}{k}\binom{2n-2k}{n+a}$

(b) $\sum_k \binom{x}{k}\binom{y}{n-k}$

(c) $\sum_k k\binom{2n+1}{2k+1}$

(d) $\sum_{k=0}^{n} \binom{n+k}{k}2^{-k}$ (Careful— watch the limits of the sum.)

(e) $\sum_k (-1)^k \binom{n-k}{k}2^{n-2k}$

2. Find recurrence formulas for the Legendre polynomials

$$P_n(x) = 2^{-n}\sum_k (-1)^k \binom{2n-2k}{n-k}\binom{n-k}{k}x^{n-2k},$$

using the creative telescoping algorithm.

3. The number $f(n)$ of involutions of n letters is given by

$$f(n) = \sum_k \frac{n!}{k!2^k(n-2k)!}.$$

Find the recurrence that is satisfied by $f(n)$, using the creative telescoping algorithm (see Example 8.4.3 on page 159).

4. Prove the identity

$$\sum_{k\le 2n}\binom{k}{n}x^k = \left(\frac{x}{1-x}\right)^n\left\{1 + x(1-2x)\sum_{j=0}^{n-1}\binom{2j+1}{j}(x(1-x))^j\right\}.$$

Hint: Use creative telescoping to find a recurrence for

$$F(n,k) := \binom{2n-k}{n}x^k.$$

Then sum over $k \ge 0$ to find a recurrence for the polynomials

$$\phi_n(x) := \sum_k F(n,k).$$

5. You are stranded on a desert island with only your laptop computer and Zeilberger's algorithm. Your rescue depends upon your being able to execute Gosper's algorithm within the next 30 seconds. What will you do?

Chapter 7

The WZ Phenomenon

7.1 Introduction

We come now to an amazingly short method for certifying the truth of combinatorial identities, due to Wilf and Zeilberger [WZ90a]. With this method, the proof certificate for an identity contains just a single rational function. That's it. Thus we will be able, for instance, to give proofs of every identity in the hypergeometric database of Chapter 3, and each proof will consist of just a certain rational function $R(n, k)$.

To put the matter in better perspective, here is a brief comparison of the WZ algorithm with the Zeilberger–Petkovšek (Z–P) route to the proofs of identities. If we are starting with an unknown hypergeometric sum and we want to know if it can be "done" in some closed form,[1] then the WZ method cannot help at all, while, on the other hand, the use of the Zeilberger algorithm, if necessary followed by Petkovšek's algorithm, will *with certainty* give a full answer to the question. So if you want to "do" an unknown sum of factorials and powers etc., the Z–P algorithms are a guaranteed route to the answer.

What the WZ algorithm can do is the following:

- It can provide extremely succinct proofs of *known* identities.

- It allows us to discover *new identities* whenever it succeeds in finding a proof certificate for a known identity.

Hence the objectives of this method and of the others are somewhat different.

---

[1] See page 145.

Suppose we want to prove an identity $\sum_k F(n,k) = r(n)$. First, if the right side, $r(n)$, is nonzero, divide through by that right hand side, and write the identity that is to be proved as

$$\sum_k \left\{ \frac{F(n,k)}{r(n)} \right\} = 1.$$

That being done, we might as well just think of $F(n,k)/r(n)$ as having been the original summand. In other words, we can now assume, without loss of generality, that the identity that we're trying to prove is

$$\sum_k F(n,k) = \text{const.} \qquad (7.1.1)$$

Let's call the left hand side $f(n)$. So $f(n) \overset{def}{=} \sum_k F(n,k)$, and we're trying to prove that $f(n) = \text{const.}$ for all n. One way to show that a function f is constant is to show that $f(n+1) - f(n) = 0$ for all n. That would certainly do it.

A good way to certify the fact that $f(n+1) - f(n) = 0$ for all n would be to display a function $G(n,k)$ such that

$$F(n+1,k) - F(n,k) = G(n,k+1) - G(n,k), \qquad (7.1.2)$$

for then we would simply sum (7.1.2) over all integers k to find that, under suitable hypotheses, indeed $f(n+1) - f(n)$ is always 0.

A pair of functions (F,G) that satisfy (7.1.2) is called a *WZ pair*.

Example 7.1.1. Let's convince ourselves that the function

$$f(n) = \sum_k \frac{\binom{n}{k}^2}{\binom{2n}{n}} \qquad (7.1.3)$$

is always equal to 1 for $n \geq 0$. To do that we state that (7.1.2) is indeed true, where F is the summand (not the sum; the *summand*) in (7.1.3), and $G = RF$, where $R(n,k)$ is the rational function

$$R(n,k) = -\frac{k^2(3n - 2k + 3)}{2(2n+1)(n-k+1)^2}. \qquad (7.1.4)$$

As usual, when confronted with such things, we suppress the natural where-on-earth-did-that-come-from reaction, and we confine ourselves to verifying the certificate. To verify it we check that if $R(n,k)$ is given by (7.1.4), we multiply R by $F(n,k)$, the summand in (7.1.3), to get a certain function $G(n,k)$. We then take that function and check that (7.1.2) is satisfied, and we're all finished. $\qquad \square$

Now let's discuss where on earth it came from. Begin with your summand $F(n, k)$, from (7.1.1). Now form the difference $D = F(n + 1, k) - F(n, k)$, and collect and simplify it as much as you can (after all, it's only a rational function times $F(n, k)$). This difference D depends on n and k and the various parameters that may have been in your summand. Let's think of n, for the time being, as one of the parameters, so we won't explicitly name it as one of the variables that D depends on. That means that D is a function of k alone, so call it $D(k)$.

Take $D(k)$ and give it to Gosper's algorithm. If possible, Gosper's algorithm will produce a function $g(k)$ such that $D(k) = g(k + 1) - g(k)$. That function $g(k)$ will of course have the parameter n in it, so let's rename $g(k)$ to $G(n, k)$. This $G(n, k)$ does exactly what we wanted it to do, namely it is the WZ-mate for F in equation (7.1.2). Furthermore, by equation (5.2.1), G/F is a rational function.

There's just one problem.

There is no assurance that Gosper's algorithm will find a g. After all, Gosper's algorithm might just return "No such g exists." There is no theorem that says that such a g must exist. In fact, there are cases where the identity $\sum_k F(n, k) = \text{const.}$ is true, but the g really doesn't exist.

Despite that, the simple fact remains that among hundreds of hypergeometric identities for which it has been tried, the WZ certification does indeed work for all but a handful of them.

Of course the earlier results *always* work. So for every proper hypergeometric term $F(n, k)$ we always have a ("telescoping") certification of (7.1.1) that looks like

$$\sum_{j=0}^{J} a_j(n) F(n + j, k) = G(n, k + 1) - G(n, k). \qquad (7.1.5)$$

The observed fact is that 99% of the time, this reduces to just two terms on the left and becomes the WZ equation (7.1.2), if we take the precaution of first dividing the summand by the right hand side of the identity, if that right hand side was not already zero.

So the thing to remember is always to give the WZ phenomenon a *chance* to happen by dividing through your identity by its right hand side if necessary. Then the identity will be in the standard form (7.1.1). If one now applies Zeilberger's telescoping algorithm to the identity *in standard form*, then there is a superb chance that it will give us a recurrence, as output, that is in the WZ form (7.1.2).

When this happens it is important that it be recognized, for several reasons that we are about to discuss. One reason is a metaphysical one. When the WZ equation (7.1.2) holds, there is complete symmetry between the indices n and k, especially for terminating identities, which previously had seemed to be playing seemingly different roles. The revelation of symmetry in nature has always been one of the main objectives of science. We will explore some of the consequences of this symmetry in this chapter.

How to prove an identity from its WZ certificate

To prove an identity $\sum_k f(n,k) = r(n)$ from its WZ certificate $R(n,k)$:

- If $r(n) \neq 0$, then put $F(n,k) := f(n,k)/r(n)$, else put $F(n,k) := f(n,k)$. Let $G(n,k) := R(n,k)F(n,k)$.

- Verify that equation (7.1.2) is true. To do this, write it all out, divide out all of the factorials, and verify the resulting polynomial identity.

- Verify that the given identity is true for one value of n.   □

The theorem that underlies these procedures makes use of these hypotheses:

- (F1) For each integer k, the limit

$$f_k = \lim_{n \to \infty} F(n,k) \qquad (7.1.6)$$

  exists and is finite.

- (G1) For each integer $n \geq 0$, we have

$$\lim_{k \to \pm\infty} G(n,k) = 0.$$

- (G2) We have $\lim_{L \to \infty} \sum_{n \geq 0} G(n,-L) = 0$.

How to find the WZ certificate of an identity

To certify an identity that is in the form $\sum_k f(n,k) = r(n)$:

- If $r(n) \neq 0$ then let $F(n,k) := f(n,k)/r(n)$, else let $F(n,k) := f(n,k)$.

- Let $f(k) := F(n+1,k) - F(n,k)$. Input $f(k)$ to Gosper's algorithm. If that algorithm fails then this one does too.

- Otherwise, the output $G(n,k)$ of Gosper's algorithm is the WZ mate of F. The rational function $R(n,k) := G(n,k)/F(n,k)$ is the WZ certificate of the identity $\sum_k F(n,k) = \text{const.}$ □

The theorem itself is the following.

Theorem 7.1.1 *[WZ90a] Let (F, G) satisfy equation (7.1.2). If (G1) holds, then*

$$\sum_k F(n, k) = const \qquad (n = 0, 1, 2, \ldots). \qquad (7.1.7)$$

If (F1) and (G2) both hold, then we have the companion identity

$$\sum_{n \geq 0} G(n, k) = \sum_{j \leq k-1} (f_j - F(0, j)), \qquad (7.1.8)$$

where f is defined by (7.1.6).

The theorem not only validates the certification procedure, it shows that we get a free new identity every time we use the method. The new identity is the companion identity (7.1.8). Roughly the reason that it is there is that the functions F, G play very symmetrical roles in the WZ equation (7.1.2). If they are so symmetric, why should there be an identity associated with F and not with G? Well there *is* one associated with G, and it is (7.1.8).

This is not the only *free* identity that we get from the procedure. We will discuss at least two more before we're finished. None of them appear as results of any of the earlier certification procedures that we have discussed. That is, if we use the raw two-variable recurrence for F, or the telescoping recurrence for F as our certification method, we get the advantage that they are guaranteed to work on every proper hypergeometric summand, but a disadvantage relative to the WZ method is that we get no identities that we didn't know before.

Proof. We use the symbol Δ_n for the forward difference operator on n: $\Delta_n h(n) = h(n+1) - h(n)$. Sum both sides of equation (7.1.2) from $k = -L$ to $k = K$, getting

$$\Delta_n \left\{ \sum_{k=-L}^{K} F(n, k) \right\} = \sum_{k=-L}^{K} \{\Delta_k G(n, k)\}$$
$$= G(n, K+1) - G(n, -L).$$

Now let $K, L \to \infty$ and use (G1) to find that $\Delta_n \sum_k F(n, k) = 0$, i.e., that $\sum_k F(n, k)$ is independent of $n \geq 0$, which establishes (7.1.7).

If we sum both sides of (7.1.2) from $n = 0$ to N, we get

$$F(N+1, k) - F(0, k) = \Delta_k \left\{ \sum_{n=0}^{N} G(n, k) \right\}.$$

Now let $N \to \infty$ and use (F1) to get

$$f_k - F(0, k) = \Delta_k \left\{ \sum_{n \geq 0} G(n, k) \right\}.$$

Replace k by k', sum from $k' = -L$ to $k' = k - 1$, let $L \to \infty$, and use (G2) to obtain the companion identity (7.1.8), completing the proof.   □

7.2   WZ proofs of the hypergeometric database

To illustrate the scope of the WZ method, we now present one-line proofs of every named identity in the hypergeometric database of Chapter 3.

(I) Proof of Gauss's $_2F_1$ identity. To prove Gauss's identity $\sum_k F(n, k) = 1$ where

$$F(n, k) = \frac{(n + k)! \, (b + k)! \, (c - n - 1)! \, (c - b - 1)!}{(c + k)! \, (n - 1)! \, (c - n - b - 1)! \, (k + 1)! \, (b - 1)!},$$

take

$$R(n, k) = \frac{(k + 1)(k + c)}{n(n + 1 - c)}. \quad □$$

(II) Proof of Kummer's $_2F_1$ identity. To prove that

$$_2F_1 \left[\begin{matrix} 1 - c - 2n & -2n \\ & c \end{matrix} ; -1 \right] = (-1)^n \frac{(2n)! \, (c - 1)!}{n! \, (c + n - 1)!}$$

rewrite it as $\sum_k F(n, k) = 1$ where

$$F(n, k) = (-1)^{n+k} \frac{(2n + c - 1)! \, n! \, (n + c - 1)!}{(2n + c - 1 - k)! \, (2n - k)! \, (c + k - 1)! \, k!}.$$

Then take $R(n, k)$ to be

$$\frac{k(k + c - 1)(2 + 4c + c^2 - 3k - 2ck + k^2 + 10n + 7cn - 6kn + 10n^2)}{2(2n - k + c + 1)(2n - k + c)(2n - k + 2)(2n - k + 1)}. \quad □$$

(III) Proof of Saalschütz's $_3F_2$ identity. To prove Saalschütz's $_3F_2$ identity in the form $\sum_k F(n, k) = \text{const.}$, where

$$F(n, k) = \frac{(a + k - 1)! \, (b + k - 1)! \, n! \, (n + c - a - b - k - 1)! \, (n + c - 1)!}{k! \, (n - k)! \, (k + c - 1)! \, (n + c - a - 1)! \, (n + c - b - 1)!},$$

take

$$R(n,k) = \frac{k(-1+c+k)(a+b-c+k-n)}{(b-c-n)(-a+c+n)(1-k+n)}. \quad \square$$

(IV) Proof of Dixon's identity. To prove that

$$\sum_k (-1)^k \binom{n+b}{n+k}\binom{n+c}{c+k}\binom{b+c}{b+k} = \frac{(n+b+c)!}{n!\,b!\,c!},$$

take

$$R(n,k) = \frac{(k+b)(k+c)}{2(k-n-1)(n+b+c+1)}. \quad \square$$

(V) Proof of Clausen's $_4F_3$ identity. To prove Clausen's $_4F_3$ identity in the form $\sum_k F(n,k) = 1$, where

$$F(n,k) = \phi(k)\phi(n-k)\psi(n),$$

and

$$\phi(t) = \frac{(a+t-1)!\,(b+t-1)!}{t!\,(-\frac{1}{2}+a+b+t)!},$$

$$\psi(t) = \frac{(t+a+b-\frac{1}{2})!\,t!\,(t+2a+2b-1)!}{(t+2a-1)!\,(t+a+b-1)!\,(t+2b-1)!},$$

take $R(n,k)$ to be

$$\frac{(1-2a-2b-2k)k(a-k+n)(b-k+n)(2+2a+2b-2k+3n)}{(2a+n)(a+b+n)(2b+n)(1-k+n)(1+2a+2b-2k+2n)}. \quad \square$$

(VI) Proof of Dougall's $_7F_6$ identity. To prove Dougall's $_7F_6$ identity

$$_7F_6\left[\begin{matrix} d & 1+\frac{d}{2} & d+b-a & d+c-a & 1+a-b-c & n+a & -n \\ & \frac{d}{2} & 1+a-b & 1+a-c & b+c+d-a & 1+d-a-n & 1+d+n \end{matrix}\,\Big|\,1\right]$$

$$= \frac{(d+1)_n(b)_n(c)_n(1+2a-b-c-d)_n}{(a-d)_n(1+a-b)_n(1+a-c)_n(b+c+d-a)_n},$$

take $R(n,k)$ to be

$$\frac{(a-c+k)(a-b+k)(a-b-c+k)(-1-a+b+c+d+k)(a-d-k+n)(1+a+2n)}{(-a+b+c-k)(d+2k)(a+n)(b+n)(c+n)(-1+2a-b-c-d+n)(1-k+n)}. \quad \square$$

7.3   Spinoffs from the WZ method

The main business of the WZ method is the certification of identities. In the course of getting that job done, however, it gives as dividends at least three additional kinds of new identities for each one that it certifies. These are (1) the companion identity (2) dual identities and (3) the definite-sum-made-indefinite. Here we will discuss and illustrate these three kinds of spinoffs.

The companion identity

The companion identity is equation (7.1.8). It states that

$$\sum_{n \geq 0} G(n, k) = \sum_{j \leq k-1} (f_j - F(0, j)), \tag{7.3.1}$$

in which

$$f_k \overset{def}{=} \lim_{n \to \infty} F(n, k).$$

Example 7.3.1.   Consider once more the identity

$$\sum_k \binom{n}{k}^2 = \binom{2n}{n}.$$

Here $F(n, k) = \binom{n}{k}^2 / \binom{2n}{n}$, and all f_k's are 0. To do the WZ procedure we apply Gosper's algorithm to the input $F(n + 1, k) - F(n, k)$, and it outputs

$$G(n, k) = \frac{(-3n + 2k - 3)}{2(2n + 1)} \frac{n!^2}{(k - 1)!^2 (n - k + 1)!^2 \binom{2n}{n}}.$$

If we substitute in (7.1.8) and use the fact that $F(0, j) = \delta_{0,j}$ we obtain the companion identity

$$\sum_{n \geq 0} \frac{(-3n + 2k - 3)}{2(2n + 1)} \frac{n!^2}{(k - 1)!^2 (n - k + 1)!^2 \binom{2n}{n}} = \begin{cases} 0 & \text{if } k = 0; \\ -1 & \text{if } k \geq 1, \end{cases}$$

which simplifies to

$$\sum_{n \geq 0} \frac{(3n - 2k + 1)}{(2n + 1)} \frac{\binom{n}{k}^2}{\binom{2n}{n}} = 2 \qquad (k = 0, 1, 2, \ldots). \tag{7.3.2}$$

□

Equation (7.3.2) is a new identity, in the sense that it is not immediately reducible to any known identity in the hypergeometric database. This comment is made here not so much to impress the reader with what a spectacular identity (7.3.2) is, but rather to underline the incompleteness of any fixed database. It seems that whatever finite list of general identities one may incorporate into a database will be incomplete. One can add, of course, a number of hypergeometric transformation rules, which map given identities onto other ones. These greatly extend the scope of any database, but still there is no such fixed list of identities and list of rules that will prove every preassigned identity.

Practically every small binomial coefficient identity is a special case of some known, more general, hypergeometric identity. But the other side of that coin is that virtually all of them can be certified directly by WZ certificates which yield, as a byproduct, a companion identity that might be new. Of course, not all of these companion identities will be æsthetically delightful. Many of them are quite messy. But they are mostly beyond the reach of any fixed database and searching and transforming algorithm that is known. The WZ approach provides a systematic way to find and to prove them by computer.

Another point is the following. Suppose identity x is a special case of identity X. *It does not follow that the companion of identity x is necessarily a special case of the companion of identity X.* In fact, this is usually false. A similar remark will hold for the idea of a *dual* identity, which we will treat later in this section. The implication of this fact is that one might be able to find a new identity from the companion of a special case, even though the companion of the more general identity was not new.

Example 7.3.2. Next let's take the identity (2.3.2) of Chapter 2, which was

$$\sum_k \frac{(n-i)!\,(n-j)!\,(i-1)!\,(j-1)!}{(n-1)!\,(k-1)!\,(n-i-j+k)!\,(i-k)!\,(j-k)!} = 1, \qquad (7.3.3)$$

and find its companion.

The rational function certificate was simply $R(n,k) = (k-1)/n$, so

$$G(n,k) = \frac{(n-i)!\,(n-j)!\,(i-1)!\,(j-1)!}{n!\,(k-2)!\,(n-i-j+k)!\,(i-k)!\,(j-k)!}.$$

The summands here are well defined only for $1 \le i, j \le n$. Hence, to fix ideas, suppose that $1 \le i \le j \le n$. Now we calculate the limits that are needed in the companion identity, namely

$$
\begin{aligned}
f_k &= \lim_{n \to \infty} F(n, k) \\
&= \lim_{n \to \infty} \frac{(n-i)^{n-i}(n-j)^{n-j}e^{k-1}(i-1)!\,(j-1)!}{(n-1)^{n-1}(n-i-j+k)^{n-i-j+k}(k-1)!\,(i-k)!\,(j-k)!} \\
&= \frac{(i-1)!\,(j-1)!}{(k-1)!\,(i-k)!\,(j-k)!} \lim_{n \to \infty} n^{1-k} = \begin{cases} 1, & \text{if } k = 1; \\ 0, & \text{if } k \geq 2. \end{cases}
\end{aligned}
$$

Thus $f_k = \delta_{1,k}$.

Next we take the basic WZ equation, in the form

$$
F(n+1, k') - F(n, k') = G(n, k'+1) - G(n, k'),
$$

which is valid only for $n \geq j$, and we sum it over all $n \geq j$ and $k' < k$. The result is

$$
1 - \sum_{k' \leq k-1} F(j, k') = \sum_{n \geq j} G(n, k),
$$

i.e.,

$$
\frac{(i-1)!\,(j-1)!}{(k-2)!\,(i-k)!\,(j-k)!} \sum_{n \geq j} \frac{(n-i)!\,(n-j)!}{n!\,(n-i-j+k)!} = 1 - \sum_{k' \leq k-1} \binom{j-i}{j-k'}\binom{i-1}{i-k'}.
$$

But in the last sum the only nonzero term comes from $k' = i$, so, after writing $n - i - j + k = r$ in the sum on the left, this reduces to

$$
\sum_{r=0}^{\infty} \frac{(r+j-k)!\,(r+i-k)!}{r!\,(r+i+j-k)!} = \frac{(k-2)!\,(i-k)!\,(j-k)!}{(i-1)!\,(j-1)!}, \tag{7.3.4}
$$

which is Gauss's $_2F_1$ identity.

What we have found, in this example, is that Gauss's identity is the companion of the identity (2.3.2), whose proof certificate was simply $(k-1)/n$. □

There is still more to say about the identity (7.3.3). Not only is the sum over k equal to 1, but the sum over i of the same summand is equal to n/j. We state this as

$$
\sum_i \frac{(n-i)!\,(n-j)!\,(i-1)!\,j!}{n!\,(k-1)!\,(n-i-j+k)!\,(i-k)!\,(j-k)!} = 1 \qquad (1 \leq k \leq j \leq n). \tag{7.3.5}
$$

We leave the investigation of this identity to the exercises. □

Example 7.3.3. If we begin with the full Saalschütz identity, as in eq. (3.5), then we find the companion identity

$$
{}_3F_2\left[\begin{matrix} k & c-a-b & c+k-1 \\ & c-b+k & c-a+k \end{matrix}; 1\right] = \frac{(c-b)_k(c-a)_k}{(a)_k(b)_k}\left\{1 - \frac{(c-a-b)_a}{(c-a)_a}\sum_{j=0}^{k-1}\frac{(a)_j(b)_j}{j!\,(c)_j}\right\}.
$$

This is valid for $c > a+b$, integer $k > 0$, and the sum is again nonterminating. It is interesting in that on the right side we see a partial sum of Gauss's original ${}_2F_1[1]$ identity. $\qquad\square$

Example 7.3.4. This time, let's start with Vandermonde's identity

$$
\sum_k\binom{a}{k}\binom{n}{k} = \binom{n+a}{a}.
$$

For this we find the WZ certificate

$$
R(n,k) = \frac{k^2}{(-1+k-n)(1+a+n)}
$$

and the companion identity

$$
\sum_n\frac{\binom{n}{k}}{\binom{n+a+1}{n}} = \frac{a+1}{(k+1)\binom{a}{k+1}},
$$

which is valid for integer $a > k \geq 0$. $\qquad\square$

Dual identities

The mapping from an identity to its companion is not an involution. But there is a true dual identity in the WZ theory. It is obtained as follows.

Suppose $F(n,k)$ is a summand of the form

$$
F(n,k) = x^n y^k \rho(n,k)\frac{\prod_i(a_i n + b_i k + c_i)!}{\prod_i(u_i n + v_i k + w_i)!},
$$

where ρ is a rational function of n, k. We will now exhibit a certain operation that will change F into a different hypergeometric term and will change its WZ mate G into a different hypergeometric term also, *but the new F, G will still be a WZ pair*. Thus we will have found a "new identity," or at any rate, a different identity from the given one.

The operation is as follows. Find any factor $(an + bk + c)!$ that appears, say, in the numerator of F. Remove that factor from the numerator of F, and place in the denominator of F a factor $(-1 - an - bk - c)!$. Then multiply F by $(-1)^{an+bk}$. By doing this to F we will have changed F to a new function, say $\tilde{F}$. Perform exactly the same operation on the WZ mate, G, of F, getting $\tilde{G}$.

We claim that $\tilde{F}$, $\tilde{G}$ are still a WZ pair.[2]

To see why, begin with the reflection formula for the gamma function,

$$\Gamma(z)\Gamma(1 - z) = \frac{\pi}{\sin \pi z}.$$

Thus we have

$$(an + bk + c)! = -\frac{\pi}{\sin\left(\pi(an + bk + c)\right)(-1 - an - bk - c)!}.$$

It follows that if we take some term F, and divide it by $(an + bk + c)!$, and then divide it by $(-1)^{an+bk}(-1 - an - bk - c)!$, then we have in effect multiplied the term F by the factor

$$\left\{ (-1)^{an+bk+1} \frac{\sin\left(an + bk + c\right)\pi}{\pi} \right\}.$$

But the remarkable thing about this latter factor is that *it is a periodic function of n and k of period 1.* Consequently, if we perform this operation

$$(an + bk + c)! \longrightarrow (-1)^{an+bk}/(-an - bk - c - 1)! \qquad (7.3.6)$$

on both F and G, then the basic WZ equation

$$F(n + 1, k) - F(n, k) = G(n, k + 1) - G(n, k)$$

will still hold between the new functions $\tilde{F}, \tilde{G}$, since F, G will merely have been multiplied by functions of period 1. □

Thus we can carry out this operation on any factorial factor in the numerators, putting a different factorial factor in the denominators, or vice-versa, and we can perform this operation repeatedly, on different factorial factors that appear in F and G, all the while preserving the WZ pair relationship. Hence we can manufacture many dual identities from the original one. Some of these may be uninteresting, some of them may be nonterminating and divergent, but some may be interesting, too.

---

[2]This result is due to Wilf and Zeilberger [WZ90a], but the following proof is from Gessel [Gess95].

Note that (check this!) the mapping

$$(F(n,k), G(n,k)) \longrightarrow (G(-k-1,-n), F(-k,-n-1)) \qquad (7.3.7)$$

also maps WZ pairs to WZ pairs (see Section 7.4 for more such transformations).

Example 7.3.5. We return to the sum of the squares of the binomial coefficients identity of Example 7.1.1, where the WZ pair was

$$F(n,k) = \frac{\binom{n}{k}^2}{\binom{2n}{n}}; \quad G(n,k) = \frac{k^2(-3+2k-3n)}{2(2n+1)(n-k+1)^2} \frac{\binom{n}{k}^2}{\binom{2n}{n}}.$$

We apply the mapping (7.3.6) to all of the factors in F except for the factor $(n-k)!^2$, and discover that the original pair F, G has been mapped to the "shadow" pair

$$\overline{F}(n,k) = \frac{(-2n-1)!\,(-k-1)!^2}{(-n-1)!^4(n-k)!^2}; \quad \overline{G}(n,k) = (3-2k+3n)\binom{-k}{-n-1}^2\binom{-2n-2}{-n-1}/2.$$

Finally we change variables as in (7.3.7) to pass to a dual pair,

$$F'(n,k) = \frac{-3k+2n}{2}\binom{n}{k}^2\binom{2k}{k}; \quad G'(n,k) = \frac{k}{2}\binom{n}{k-1}^2\binom{2k}{k}.$$

We now have a formal WZ pair. We check that the hypotheses (G1), (G2) of the theorem are satisfied, and then we know that $\sum_k F'(n,k)$ is independent of $n \geq 0$. Since it is 0 when $n = 0$ we have a proof of the dual identity

$$\sum_k (3k-2n)\binom{n}{k}^2\binom{2k}{k} = 0 \qquad (n=0,1,\dots).$$

Again, while this identity is hardly spectacular in itself, it is new in that there is no immediate, algorithmic way to deduce it from the standard hypergeometric database (though we certainly would not want to challenge human mathematicians to give independent proofs of it!). □

We summarize a few other instances of duality. A dual of $\sum_k \binom{n}{k} = 2^n$ is

$$\sum_k (-1)^{n+k}\binom{n}{k}2^k = 1 \quad (n=0,1,\dots).$$

The Saalschütz identity is self-dual.

A dual of Dixon's identity is a special case of Saalschütz's. Note that since dualization is symmetric, it follows that we can prove Dixon's identity from Saalschütz's by dualization of a special case.

A dual of Vandermonde's identity

$$\sum_k \binom{a}{k}\binom{n}{k} = \binom{n+a}{a}$$

is

$$\sum_k (-1)^{n+k} \binom{n}{k}\binom{k+b}{k} = \binom{b}{n},$$

which is a special case of Gauss's $_2F_1$ identity. Again, Vandermonde's identity can thereby be proved from Gauss's, by specializing and dualizing.

The process of dualization does not in general commute with specialization. Thus we can imagine beginning with some identity, passing to some special case, dualizing, passing to some special case, etc., thereby generating a whole chain of identities from a given one. If the original identity has a large number of free parameters, like Dougall's $_7F_6$ for instance, then the chain might be fairly long. Whether interesting new identities can be found this way is not known.

The definite sum made indefinite

The following observation of Zeilberger generates an interesting family of identities and is of help in finding asymptotic estimates of hypergeometric sums.

Imagine a sum $\sum_k F(n,k) = 1$ of the type that we have been considering, in which the support of F properly contains the interval $[0, n]$. Now consider the restricted sum

$$h(n) := \sum_{k=0}^{n} F(n,k).$$

A prototype of this situation is a sum like $\sum_{k=0}^{n} \binom{3n}{k}$.

Suppose now that F has a WZ mate G, so that

$$F(n+1,k) - F(n,k) = G(n,k+1) - G(n,k).$$

Sum this equation over $k \leq n$ to obtain

$$h(n+1) - F(n+1,n+1) - h(n) = G(n,n+1).$$

Next, replace n by j and sum over $j = 0, \ldots, n-1$ to get

$$h(n) = h(0) + \sum_{j=1}^{n} (F(j,j) + G(j-1,j)),$$

which is to say that

$$\sum_{k=0}^{n} F(n,k) = F(0,0) + \sum_{j=1}^{n}(F(j,j) + G(j-1,j)). \qquad (7.3.8)$$

We notice that on the right side we have a sum in which the running index n does not appear under the summation sign.

Example 7.3.6. Take $F(n,k) = \binom{3n}{k}/8^n$, so that $\sum_k F(n,k) = 1$. The WZ mate of F is

$$G(n,k) = \frac{(-32 + 22k - 4k^2 - 93n + 30kn - 63n^2)\binom{3n}{k-1}}{(3n-k+3)(3n-k+2)} \frac{}{8^{n+1}}.$$

We find that (be sure to use a computer algebra package to do things like this!)

$$F(j,j) + G(j-1,j) = \frac{2(2-j-5j^2)}{3(3j-1)(3j-2)} 8^{-j}\binom{3j}{j}.$$

Hence (7.3.8) reads as

$$8^{-n}\sum_{k=0}^{n}\binom{3n}{k} = 1 + \frac{2}{3}\sum_{j=1}^{n}\frac{(2-j-5j^2)}{(3j-1)(3j-2)} 8^{-j}\binom{3j}{j}. \qquad (7.3.9)$$

We mention two applications of this idea, to asymptotics and to speeding up table-making.

We can find some asymptotic information as follows. It is easy to see that the left side of (7.3.9) approaches 0 as $n \to \infty$. So the sum on the right, with n replaced by ∞, is 0. Thus we have

$$\frac{2}{3}\sum_{j=1}^{\infty}\frac{(5j^2+j-2)}{(3j-1)(3j-2)} 8^{-j}\binom{3j}{j} = 1,$$

an identity that is perhaps not instantly obvious by inspection, and therefore (7.3.9) can be rewritten as

$$8^{-n}\sum_{k=0}^{n}\binom{3n}{k} = \frac{2}{3}\sum_{j=n+1}^{\infty}\frac{(5j^2+j-2)}{(3j-1)(3j-2)} 8^{-j}\binom{3j}{j}.$$

On the right side we replace j by $n+k$, factor out $8^{-n}\binom{3n}{n}$, and divide each term of the sum by that factor. We then have, as $n \to \infty$,

$$8^{-n} \sum_{k=0}^{n} \binom{3n}{k} \sim 8^{-n} \binom{3n}{n} \frac{2}{3} \sum_{k=1}^{\infty} \frac{5}{9} 8^{-k} \left(\frac{27}{4} \right)^{k}$$

$$= 2 \cdot 8^{-n} \binom{3n}{n}.$$

So the sum in question is asymptotic to twice its last term.   □

One additional application was pointed out by Zeilberger in [Zeil96b]. Suppose we want to make a table of the left hand side of (7.3.8) for $n = 1, \ldots, N$, say. If we compute directly from the left side, we will have to do $O(N)$ calculations for each n, so $O(N^2)$ all together. On the right side, however, a single new term is computed for each n, which makes all N values computable in time that is *linear* in N.

7.4   Discovering new hypergeometric identities

In this section we describe an approach due to Ira Gessel [Gess95], that uses the WZ method in yet another way, and which results in a shower of new identities. Gessel's approach is as follows.

1. Restrict attention to terminating identities. That is, assume that the summand $F(n, k)$ vanishes except for k in some compact support, i.e., a finite interval which, of course, depends on n.

2. By a WZ function, we mean any summand $F(n, k)$ for which $\sum_k F(n, k) = 1$ and which has a WZ mate $G(n, k)$. The summand may depend on parameters $a, b, c, \ldots$, and if we want to exhibit these explicitly we may write the WZ function as $F(a, b, c, \ldots, k)$. Clearly, if $F(a, b, c, \ldots, k)$ is a WZ function then so is $F(a + u_1 n, b + u_2 n, c + u_3 n, \ldots, k)$, since this simply amounts to changing the names of the free parameters. Note how this idea puts all free parameters on an equal footing, instead of singling out one of them *ab initio*, as n.

3. If $(F(n, k), G(n, k))$ is a WZ pair, then so is $(G(k, n), F(k, n))$, although the sum $\sum_k G(k, n)$ may not be terminating even though $\sum_k F(n, k)$ is.

4. If $F(n, k)$ is a WZ function, then so is $F(n + \alpha, k + \beta)$, for all α, β, and often one can choose α and β so as to make $\sum_k G(k + \beta, n + \alpha)$ terminate.

5. If $F(n, k)$ is a WZ function, then so are $F(-n, k)$ and $F(n, -k)$, and these offer further possibilities for generating new identities.

As an illustration of this approach, let's show how to find a hypergeometric identity by beginning with Dixon's identity, whose WZ function can be taken in the form

$$F_1 = (-1)^k \frac{(a+b)!\,(a+c)!\,(b+c)!\,a!\,b!\,c!}{(a+k)!\,(b-k)!\,(c+k)!\,(a-k)!\,(b+k)!\,(c-k)!\,(a+b+c)!}.$$

At the moment, the running variable n is not present. Since F_1 is a WZ function whatever the free parameters a, b, c might be, we can replace a, b, c by various functions of n, as we please. We might replace (a, b, c) by $(a+n, b-2n, c+3n)$, for instance. Just to keep things simple, let's replace a by $a+n$ and b by $b-n$, to get the WZ function $F_2(n,k)$ as

$$\frac{(-1)^k\,(a+b)!\,c!\,(b-n)!\,(b+c-n)!\,(a+n)!\,(a+c+n)!}{(a+b+c)!\,(c-k)!\,(c+k)!\,(b-k-n)!\,(b+k-n)!\,(a-k+n)!\,(a+k+n)!}.$$

Now it's time to call the WZ algorithm. So we form $F_2(n+1,k) - F_2(n,k)$, and input it to Gosper's algorithm. It returns the WZ mate of F_2, $G_2(n,k)$, as

$$\frac{(-1)^k\,(-1-a+b-2n)\,(-1+b-n)!\,(-1+b+c-n)!\,(a+n)!\,(a+c+n)!}{(-1+c-k)!\,(c+k)!\,(-1+b-k-n)!\,(b+k-n)!\,(a-k+n)!\,(1+a+k+n)!},$$

in which we have now dropped all factors that are independent of both n and k.

Now $G_2(k,n)$ will serve as a new WZ function $F_3(n,k)$, which is

$$\frac{(-1)^n\,(-1-a+b-2k)\,(-1+b-k)!\,(-1+b+c-k)!\,(a+k)!\,(a+c+k)!}{(-1+c-n)!\,(-1+b-k-n)!\,(a+k-n)!\,(c+n)!\,(b-k+n)!\,(1+a+k+n)!}.$$

As this F_3 now stands, the sum $\sum_k F_3(n,k)$ does not terminate. There are many ways to make it terminate, however. For instance, instead of F_3 we can use $F_4(n,k) = F_3(n,k+n-a)$ for our WZ function. If we do that we would certainly have $\sum_k F_4(n,k) = const$, and the "*const*" can be evaluated at any particular value of n. In this case, it turns out to be zero.

Hence we have found that $\sum_k F_4(n,k) = 0$, which can be written in the hypergeometric form

$$_5F_4\left[\begin{array}{c} -a-b, n+1, n+c+1, 2n-a-b+1, n+\frac{3-a-b}{2} \\ n-a-b-c+1, n-a-b+1, 2n+2, n+\frac{1-a-b}{2} \end{array}; 1\right] = 0. \qquad (7.4.1)$$

This is a "new" hypergeometric identity, at least in the sense that it does not live in any of the extensive databases of such identities that are available to us. We have previously discussed the fact that the word "new" is somewhat elusive. The identity

(7.4.1) might be obtainable by applying some transformation rule to some known identity, in which case it would not be "really new." Failing that, there are surely some human mathematicians who would be able to prove it by some very short application of known results, so in that extended sense it is certainly not new. But the procedure by which we found it is quite automatic. The various decisions that were made above, about how to introduce the parameter n, and how to make sure that the sum terminates, were made arbitrarily and capriciously, but if they had been made in other ways, the result would have been other "new" hypergeometric identities.

Here are three samples of other identities that Gessel found by variations of this method. His paper contains perhaps fifty more. In each case we will state the identity, and give its proof by giving the rational function $R(n, k)$ that is its WZ proof certificate.

$$_3F_2\left[\begin{array}{c} -3n, \frac{2}{3} - c, 3n + 2 \\ \frac{3}{2}, 1 - 3c \end{array} ; \frac{3}{4}\right] = \frac{(c + \frac{2}{3})_n(\frac{1}{3})_n}{(1 - c)_n(\frac{4}{3})_n},$$

$$R(n, k) = \frac{2(5 + 6n)(k - 3c)k(2k + 1)}{(3c + 3n + 2)(k - 3n - 3)(k - 3n - 2)(k - 3n - 1)}.$$

$$_3F_2\left[\begin{array}{c} -3b, \frac{-3n}{2}, \frac{1}{2} - \frac{3n}{2} \\ -3n, \frac{2}{3} - b - n \end{array} ; \frac{4}{3}\right] = \frac{(\frac{1}{3} - b)_n}{(\frac{1}{3} + b)_n},$$

$$R(n, k) = \frac{k(k - 3n - 1)(3k - 3n - 3b - 1)(3k - 5 - 6n)}{(2k - 3n - 1)(2k - 3n - 2)(2k - 3n - 3)(3b - 3n - 1)}.$$

$$_4F_3\left[\begin{array}{c} \frac{3}{2} + \frac{n}{5}, \frac{2}{3}, -n, 2n + 2 \\ n + \frac{11}{6}, \frac{4}{3}, \frac{n}{5} + \frac{1}{2} \end{array} ; \frac{2}{27}\right] = \frac{(\frac{5}{2})_n(\frac{11}{6})_n}{(\frac{3}{2})_n(\frac{7}{2})_n},$$

$$R(n, k) = \frac{9}{2}\frac{k(3k + 1)}{(k - n - 1)(2n + 10k + 5)}.$$

The ease with which such impressive identities can be manufactured shows again the inadequacy of relying solely on some fixed database of identities and underscores the flexibility and comprehensiveness of a computer-based algorithmic approach.

7.5   Software for the WZ method

In this section we will discuss how to use the programs in this book to implement the WZ method, first in Mathematica, and then in Maple.

In Mathematica one would use the program for Gosper's algorithm (see Appendix A) plus a few extra instructions. If we assume that the GosperSum program has already been read in, then the following Mathematica program will find the WZ mate and certificate $R(n, k)$:

```
(* WZ::usage="WZ[f,n,k] yields the WZ certificate of f[n,k]. Here
    the input f is an expression, not a function. If R denotes the
    rational function output by this routine, then define g[n,k] to
    be R f[n,k], to obtain a WZ pair (f,g), i.e., a pair that
    satisfies f[n+1,k]-f[n,k]=g[n,k+1]-g[n,k]" *)
WZ[f_, n_, k_] :=Module[{k1, df, t, r, g},
        df = -f + (f /. {n -> n + 1});
        t = GosperSum[df, {k, 0, k1}];
        r = FactorialSimplify[(t /. {k1 -> k-1})/f];
        g = FactorialSimplify[r f];
        Print["The rational function R(n,k) is  ",r];
        Print["The WZ mate G(n,k) is    ",g];
        Return[]
            ];
```

Example 7.5.1. To find the WZ proof of the identity

$$\sum_k \frac{2^{k+1}(k+1)(2n-k-2)!\,n!}{(n-k-1)!\,(2n)!} = 1,$$

we type

```
f=2^(k+1) (k+1) (2n-k-2)! n!/((n-k-1)! (2n)!)
WZ[f,n,k]
```

The program responds with

```
The rational function R(n,k) is k/(2(-1+k-n))
The WZ mate G(n,k) is -n!/(2^(n+1)(-1+k)! (1-k+n)!)
```

□

Example 7.5.2. Let's find the companion identity of the binomial coefficient identity

$$\sum_k k\binom{n+1}{k}\binom{x}{k} = (n+1)\binom{x+n}{n+1}. \tag{7.5.1}$$

To do that we first input to the program above the request

`WZ[k Binomial[n+1,k] Binomial[x,k]/((n+1) Binomial[n+x,n+1]),n,k]`

We are told that the rational function $R(n,k)$ is

$$-\frac{k(k-1)}{(n-k+2)(n+x+1)},$$

and that the WZ mate is

$$G(n,k) = -\frac{(k-1)(n+1)!^2 x!^2}{(n+1)x(k-1)!^2(n-k+2)!\,(x-k)!\,(x+n+1)!},$$

i.e., that

$$G(n,k) = -\frac{k(k-1)}{x(n+1)}\frac{\binom{n+1}{k-1}\binom{x}{k}}{\binom{x+n+1}{x}}.$$

To find the companion identity we must first compute the limits f_k, of (7.1.6). We find that

$$f_k = \lim_{n\to\infty} F(n,k) = \lim_{n\to\infty} \frac{k\binom{n+1}{k}\binom{x}{k}}{(n+1)\binom{x+n}{n+1}} = 0 \quad (k < x),$$

and also $F(0,j) = \delta_{j,1}$. Hence the general companion identity (7.1.8) becomes, in this case,

$$-\frac{k(k-1)}{x}\binom{x}{k}\sum_{n\geq 0}\frac{\binom{n+1}{k-1}}{(n+1)\binom{x+n+1}{x}} = \sum_{j\leq k-1}(0-\delta_{j,1}),$$

which can be tidied up and put in the form

$$\sum_{n\geq 0}\frac{n!\,(n+1)!}{(n-k+2)!\,(x+n+1)!} = \frac{(k-2)!}{k\binom{x}{k}(x-1)!} \quad (k\geq 2;\ x>k).$$

This is the desired companion, and it is a nonterminating special case of Gauss's $_2F_1$ identity.   □

Let's try the same thing in Maple. The program of choice is now the creative telescoping algorithm of Chapter 6. It can be used to find WZ proof certificates quite easily. First we write the identity under consideration in the standard form $\sum_k f(n,k) = 1$. Next we call program ct from the EKHAD package, with the call `ct(f,1,k,n,N);`, thereby asking it to look for a recurrence of ORDER:=1.[3]

Example 7.5.3. We illustrate by finding, in Maple, the WZ proof of the identity $\sum_k \binom{n}{k}^2 = \binom{2n}{n}$. First, as outlined above, we write the sum in the standard form

---

[3]See also Proposition 8.1.1 on page 147.

$\sum_k f(n, k) = 1$, where

$$f(n, k) := \frac{n!^4}{k!^2 (n - k)!^2 (2n)!}.$$

Next we execute the instruction $\texttt{ct(f(n,k),1,k,n,N)}$, and program $\texttt{ct}$ returns the pair

$$N - 1, \quad -\frac{(3n + 3 - 2k)k^2}{2(n + 1 - k)^2 (2n + 1)}.$$

These two items represent, respectively, the operator in n, N which appears on the left side of the creative telescoping equation, and the rational function $R(n, k)$ which converts the f into the g.

More generally, if the program returns a pair $\Omega(N, n), R(n, k)$, it means that the input summand $F(n, k)$ satisfies the telescoping recurrence

$$\Omega(N, n)F(n, k) = G(n, k + 1) - G(n, k), \tag{7.5.2}$$

where $G(n, k) = R(n, k)F(n, k)$. Hence, in the present example, the program is telling us that $(N - 1)F(n, k) = G(n, k + 1) - G(n, k)$, which is exactly the WZ equation. $\qquad\qquad\qquad\qquad\qquad\qquad\qquad\qquad\qquad\qquad\qquad\square$

Thus the program outputs the WZ mate $G(n, k)$ as well as the rational function certificate $R(n, k)$.

7.6   Exercises

1. Suppose $\sum_k F(n, k) = 1$ and that $F(n, k)$ satisfies the telescoped recurrence (7.5.2) in which the operator Ω is of order higher than the first, but has a *left* factor of $N - 1$. That is $\Omega = (N - 1)\Omega'$. Then $\Omega'F$ is the first member of a WZ pair. Investigate whether $N - 1$ is or is not a left factor in some instances where the creative telescoping algorithm does not find a first order recurrence.

2. Find the WZ proof of the identity (7.3.5). Then find the companion identity and relate it to known hypergeometric identities.

3. In all parts of this problem, $F(n, k)$ will be the summand of (7.3.3).

   (a) Show that all of the following are true ("Δ" is the forward difference operator w.r.t. its subscript):

   $$\Delta_n F = \Delta_k \left(\frac{k - 1}{n} F \right)$$

$$\Delta_j\left(\frac{j}{n}F\right) = \Delta_i\left(\frac{(k-i)(n-i+1)j}{(j-n)(j+1-k)n}F\right)$$

$$\Delta_n\left(\frac{j}{n}F\right) = \Delta_i\left(\frac{(k-i)(n-i+1)j}{(n+1)(n+1+k-i-j)n}F\right)$$

(b) In each of the cases above, find the companion identity and relate it to the hypergeometric database.

Chapter 8

Algorithm Hyper

8.1 Introduction

If you want to evaluate a given sum in closed form, so far the tools that have
been described in this book have enabled you to find a recurrence relation with
polynomial coefficients that your sum satisfies. If that recurrence is of order 1
then you are finished; you have found the desired closed form for your sum, as a
single hypergeometric term. If, on the other hand, the recurrence is of order ≥ 2
then there is more work to do. How can we recognize when such a recurrence has
hypergeometric solutions, and how can we find all of them?

In this chapter we discuss the question of how to recognize when a given recurrence relation with polynomial coefficients has a closed form solution. We first take
the opportunity to define the term "closed form."[1]

Definition 8.1.1 *A function $f(n)$ is said to be of closed form if it is equal to a
linear combination of a fixed number, r, say, of hypergeometric terms. The number
r must be an absolute constant, i.e., it must be independent of all variables and
parameters of the problem.*   □

Take a definite sum of the form $f(n) = \sum_k F(n,k)$ where the summand $F(n,k)$
is hypergeometric in both its arguments. Does this sum have a closed form? The
material of this chapter, taken together with the algorithm of Chapter 6, provides
a *complete* algorithmic solution of this problem.

To answer the question, we first run the creative telescoping algorithm of Chapter 6 on $F(n,k)$. It produces a recurrence satisfied by $f(n)$. If this recurrence is

---

[1] We are really defining *hypergeometric closed form.*

first-order then the answer is "yes," and we have found the desired closed form. But what if the recurrence is of order two or more? Well, then we don't know!

Example 8.1.1. Consider the sum $f(n) = \sum_k \binom{3k+1}{k}\binom{3n-3k}{n-k}/(3k+1)$. Creative telescoping produces a second-order recurrence for this sum:

```
In[1]:= <<zb_alg.m
Out[1]= Peter Paule and Markus Schorn's implementation loaded...
In[2]:= Zb[Binomial[3k+1,k] Binomial[3(n-k),n-k]/(3k+1), {k,0,n}, n, 1]
Out[2]= Try higher order
In[3]:= Zb[Binomial[3k+1,k] Binomial[3(n-k),n-k]/(3k+1), {k,0,n}, n, 2]
Out[3]= {-81 (1 + n) (2 + 3 n) (4 + 3 n) SUM[n] +
>       12 (3 + 2 n) (22 + 27 n + 9 n^2 ) SUM[1 + n] -
>       4 (2 + n) (3 + 2 n) (5 + 2 n) SUM[2 + n] == 0}
```

But browsing through a list of binomial coefficient identities (such as the one in [GKP89]), we encounter the identity

$$\sum_k \binom{tk+r}{k}\binom{tn-tk+s}{n-k}\frac{r}{tk+r} = \binom{tn+r+s}{n}. \tag{8.1.1}$$

When $t = 3$, $r = 1$, $s = 0$, this identity specializes to

$$\sum_k \binom{3k+1}{k}\binom{3n-3k}{n-k}\frac{1}{3k+1} = \binom{3n+1}{n}, \tag{8.1.2}$$

implying that our $f(n)$ is nevertheless a hypergeometric term! Our knowledge of recurrence Out[3] satisfied by $f(n)$ makes it easy to *verify* (8.1.2) independently:

```
In[4]:= FactorialSimplify[Out[3][[1,1]] /. SUM[n_] -> Binomial[3n+1,n]]
Out[4]= 0
```

Since $f(n)$ agrees with $\binom{3n+1}{n}$ for $n = 0$ and $n = 1$, it follows that indeed $f(n) = \binom{3n+1}{n}$.

Note that the summand in (8.1.1) is not hypergeometric in k or n when t is a variable. But for every fixed integer t, it is proper hypergeometric in all the remaining variables, and so is the right hand side in (8.1.1).

Let's keep $r = 1$, $s = 0$, and see what happens when $t = 4$:

```
In[5]:= Zb[Binomial[4k+1,k] Binomial[4(n-k),n-k]/(4k+1), {k,0,n}, n, 1]
Out[5]= Try higher order
In[6]:= Zb[Binomial[4k+1,k] Binomial[4(n-k),n-k]/(4k+1), {k,0,n}, n, 2]
Out[6]= Try higher order
In[7]:= Zb[Binomial[4k+1,k] Binomial[4(n-k),n-k]/(4k+1), {k,0,n}, n, 3]
```

```
Out[7]= Try higher order
In[8]:= Zb[Binomial[4k+1,k] Binomial[4(n-k),n-k]/(4k+1), {k,0,n}, n, 4]
Out[8]= {4194304 (1 + n) (2 + n) (1 + 2 n) (3 + 2 n) (3 + 4 n) (5 + 4 n)
>        (7 + 4 n) (9 + 4 n) SUM[n] -
>        73728 (2 + n) (3 + 2 n) (7 + 4 n) (9 + 4 n)
>        (10391 + 20216 n + 15224 n^2  + 5376 n^3  + 768 n^4 ) SUM[1 + n] +
>        1728 (5 + 3 n) (7 + 3 n)
>        (4181673 + 9667056 n + 9469964 n^2  + 5043584 n^3  + 1543808 n^4  +
>          258048 n^5  + 18432 n^6 ) SUM[2 + n] -
>        432 (3 + n) (5 + 3 n) (7 + 3 n) (8 + 3 n) (10 + 3 n)
>        (15433 + 14690 n + 4896 n^2  + 576 n^3 ) SUM[3 + n] +
>        729 (3 + n) (4 + n) (5 + 3 n) (7 + 3 n) (8 + 3 n) (10 + 3 n)
>        (11 + 3 n) (13 + 3 n) SUM[4 + n] == 0}
```

This time we have a recurrence of order 4, and if we live to see the answer when $t = 5$, it will be a recurrence of order 6, even though this sum satisfies a first-order recurrence with polynomial coefficients!

In fact, we conjecture that for any nonnegative integer d, there exist integers t, r, s, such that the recurrence obtained by creative telescoping for the sum in (8.1.1) is of order d or more. □

The ability of creative telescoping to find a *first-order* recurrence when one exists is closely related to the performance of the WZ method on the corresponding identity, as the following proposition shows.

Proposition 8.1.1 *Let $F(n, k)$ be hypergeometric in both variables, and such that the sum $f(n) = \sum_k F(n, k)$ exists and is hypergeometric in n. Then creative telescoping, with input $F(n, k)$, produces a first-order recurrence for $f(n)$ if and only if the WZ method succeeds in proving that $\sum_k F(n, k) = f(n)$.*

Proof. Let $r(n) = f(n+1)/f(n)$ be the rational function representing $f(n)$. Then $f(n + 1) - r(n)f(n) = 0$. Let $\bar{F}(n, k) = F(n, k)/f(n)$. Now we have the following chain of equivalences:

Creative telescoping produces a first-order recurrence for $f(n)$

$\Longleftrightarrow$ $F(n + 1, k) - r(n)F(n, k)$ is Gosper-summable w.r.t. k

$\Longleftrightarrow$ $(F(n + 1, k) - r(n)F(n, k))/f(n + 1)$ is Gosper-summable w.r.t. k

$\Longleftrightarrow$ $\bar{F}(n + 1, k) - \bar{F}(n, k)$ is Gosper-summable w.r.t. k

$\Longleftrightarrow$ WZ method succeeds in proving that $\sum_k \bar{F}(n, k)$ is constant

$\Longleftrightarrow$ WZ method succeeds in proving the identity $\sum_k F(n, k) = f(n)$.   □

Our method of solution of the problem of definite hypergeometric summation is thus along the following lines:

1. Given a definite hypergeometric sum $f(n)$, find a recurrence satisfied by $f(n)$.

2. Find all hypergeometric solutions of this recurrence.

3. Check if any linear combination of these solutions agrees with $f(n)$, for enough consecutive values of n.

We have already shown in Chapter 6 how to perform Step 1. In this chapter we discuss linear recurrences with polynomial coefficients and give algorithms that solve them within some well-behaved class of discrete functions: polynomials, rational functions, hypergeometric terms, and d'Alembertian functions. Since no general explicit solutions of such recurrences are known, these algorithms are interesting not only in connection with identities, but also in their own right.

8.2 The ring of sequences

Let K be a field of characteristic zero. We will denote by $\mathbb{N}$ the set of nonnegative integers, and by $K^{\mathbb{N}}$ the set of all sequences $(a(n))_{n=0}^{\infty}$ whose terms belong to K. With termwise addition and multiplication, $K^{\mathbb{N}}$ is a commutative ring. It is also a K-linear space (in fact, a K-algebra) since we can multiply sequences termwise with elements of K. The field K is naturally embedded in $K^{\mathbb{N}}$ as a subring, by identifying $u \in K$ with the constant sequence $(u, u, \ldots) \in K^{\mathbb{N}}$.

The ring $K^{\mathbb{N}}$ cannot be embedded into a field since it contains zero divisors. For example, let $a = (1, 0, 1, 0, \ldots)$ and $b = (0, 1, 0, 1, \ldots)$; then

$$ab = (0, 0, 0, 0, \ldots) = 0$$

although $a, b \neq 0$. For somebody who is used to solving equations in fields this has strange consequences. For instance, the simple quadratic equation

$$x^2 = 1$$

is satisfied by any sequence with terms ± 1, hence it has a continuum of solutions!

Here we are interested not in algebraic but in recurrence equations, therefore we define the *shift operator* $N : K^{\mathbb{N}} \to K^{\mathbb{N}}$ by setting

$$N(a(0), a(1), \ldots) = (a(1), a(2), \ldots),$$

or, more compactly, $(Na)(n) = a(n + 1)$. Applying the shift operator k times, we shift the sequence k places to the left: $(N^k a)(n) = a(n + k)$.

Since $N(a + b) = Na + Nb$ and $N(\lambda a) = \lambda Na$ for all $a, b \in K^{\mathbb{N}}$ and $\lambda \in K$, the shift operator and its powers are linear operators on the K-linear space $K^{\mathbb{N}}$.

Similarly, multiplication by a fixed sequence is a linear operator on $K^{\mathbb{N}}$. Note that the set of all linear operators on $K^{\mathbb{N}}$ with addition defined pointwise and with functional composition as multiplication is a (noncommutative) ring. In particular, operators of the form

$$L = \sum_{k=0}^{r} a_k N^k,$$

where $a_k \in K^{\mathbb{N}}$, are called *linear recurrence operators* on $K^{\mathbb{N}}$. If $a_r \neq 0$ and $a_0 \neq 0$, the *order* of L is $\operatorname{ord} L = r$. A *linear recurrence equation* in $K^{\mathbb{N}}$ is an equation of the form

$$Ly = f,$$

where L is a linear recurrence operator on $K^{\mathbb{N}}$ and $f \in K^{\mathbb{N}}$. This equation is *homogeneous* if $f = 0$, and *inhomogeneous* otherwise. Note that the set of all solutions of $Ly = 0$ is a linear subspace $\operatorname{Ker} L$ of $K^{\mathbb{N}}$ (the *kernel* of L), and the set of all solutions of $Ly = f$ is an affine subspace of $K^{\mathbb{N}}$.

From the theory of ordinary differential equations we are used to the fact that a homogeneous linear differential equation of order r has r linearly independent solutions, and we expect – and desire – a similar state of affairs with recurrence equations. However, unusual things happen again.

Example 8.2.1. Let $a = (1, 0, 1, 0, \ldots)$ and $b = (0, 1, 0, 1, \ldots)$ as above. Consider the equation $L_1 y = 0$, where $L_1 = aN + b$. Rewriting this equation termwise, we have $a(n)y(n+1) + b(n)y(n) = 0$ for all n, or $y(n+1) = 0$ for n even and $y(n) = 0$ for n odd. It follows that $L_1 y = 0$ if and only if y has the form $(y(0), 0, y(2), 0, y(4), \ldots)$ where the values at even arguments are arbitrary. Thus the solution space of this first-order equation has infinite dimension!

Now consider the equation $L_2 y = 0$, where $L_2 = aN - 1$. Termwise this means that $a(n)y(n+1) - y(n) = 0$, or $y(n) = y(n+1)$ for n even and $y(n) = 0$ for n odd. It follows that $L_2 y = 0$ if and only if $y = 0$. Here we have a first-order equation with a zero-dimensional solution space! □

As the attentive reader has undoubtedly noticed, the unusual behavior in these examples stems from the fact that the sequences a and b contain infinitely many zero terms. We will be interested in linear recurrence operators with polynomial coefficients which, if nonzero, can vanish at most finitely many times. But even in this case solutions can be plentiful.

Example 8.2.2. Let $L = pN - q$ where $p(n) = (n-1)(n-4)(n-7)$ and $q(n) = n(n-3)(n-6)$. Termwise we have $y(1) = 0$, $10y(3) - 8y(2) = 0$, $y(4) = 0$, $8y(6) - 10y(5) = 0$, $y(7) = 0$, $y(n+1) = (q(n)/p(n))y(n)$ for $n \geq 8$. This yields four linearly independent solutions: $(1, 0, 0, \ldots), (0, 0, 1, 4/5, 0, 0, \ldots), (0, 0, 0, 0, 0, 1, 5/4, 0, 0, \ldots)$, and $y(n) = (n-1)(n-4)(n-7)$. Apparently, for every integer $k > 0$, a first-order equation with polynomial coefficients can have a k-dimensional solution space. $\square$

We wish to establish an algebraic setup in which $\dim \operatorname{Ker} L = \operatorname{ord} L$ for every linear recurrence operator with polynomial coefficients. Looking at the last example, we see that of the four solutions, three have only finitely many nonzero terms. This observation leads to the idea of identifying such sequences with 0, and more generally, of identifying sequences which agree from some point on. Thus an equality $a = b$ among two sequences will in fact mean

$$a(n) = b(n) \text{ a.e.},$$

where "a.e." stands for "almost everywhere" and indicates that the stated equality is valid for all but finitely many $n \in \mathbb{N}$. In particular, $a = 0$ if $a(n) = 0$ for all large enough n. We will denote the ring of sequences over K with equality taken in this "almost everywhere" sense by $\mathcal{S}(K)$.

In the next paragraph, which can be omitted at a first reading, we give a precise definition of $\mathcal{S}(K)$.

A sequence which is zero past the kth term is annihilated by N^k. The algebraic structure that will have the desired properties is therefore the quotient ring $\mathcal{S}(K) = K^{\mathbb{N}}/J$ where

$$J = \bigcup_{k=0}^{\infty} \operatorname{Ker} N^k$$

is the ideal of eventually zero sequences. Let $\varphi : K^{\mathbb{N}} \to \mathcal{S}(K)$ denote the canonical epimorphism which maps a sequence $a \in K^{\mathbb{N}}$ into its equivalence class $a + J \in \mathcal{S}(K)$. Then $\varphi N : K^{\mathbb{N}} \to \mathcal{S}(K)$ is obviously an epimorphism of rings. Since

$$\operatorname{Ker} \varphi N = (\varphi N)^{-1}(0) = N^{-1}(\varphi^{-1}(0)) = N^{-1}(J) = \bigcup_{k=1}^{\infty} \operatorname{Ker} N^k = J,$$

there is a unique automorphism E of $\mathcal{S}(K)$ such that $\varphi N = E\varphi$. We call E the *shift operator on* $\mathcal{S}(K)$. For simplicity, we will keep talking about sequences where we actually mean their corresponding equivalence classes, and will write a instead of $a + J$, and N instead of E.

Notice that a nonzero sequence $a \in \mathcal{S}(K)$ is a unit (i.e., invertible w.r.t. multiplication,) if and only if it is not a zero divisor. Namely, a is invertible iff it is eventually nonzero, and it is a zero divisor iff it contains infinitely many zero terms and infinitely many nonzero terms. For a nonzero sequence, these two properties are obviously complementary.

Now we can show that linear recurrence operators on $\mathcal{S}(K)$ have the desired properties.

Theorem 8.2.1 *Let* $L = \sum_{k=0}^{r} a_k N^k$ *be a linear recurrence operator of order* r *on* $\mathcal{S}(K)$. *If* a_r *and* a_0 *are units, then* $\dim \operatorname{Ker} L = r$.

Proof. First we show that $\dim \operatorname{Ker} L \leq r$.

Let $y_1, y_2, \ldots, y_{r+1}$ be solutions of the equation $Ly = 0$. Then there exists an $n_0 \in \mathbb{N}$ such that for all $n \geq n_0$,

$$\sum_{k=0}^{r} a_k(n) y_i(n+k) = 0, \quad \text{for } i = 1, 2, \ldots, r+1. \tag{8.2.1}$$

Let $\mathbf{y}(n) \in K^{r+1}$ be the vector with components $y_1(n), y_2(n), \ldots, y_{r+1}(n)$, for all $n \geq 0$. Denote by $\mathcal{L}_n$ the linear span of $\mathbf{y}(n), \mathbf{y}(n+1), \ldots, \mathbf{y}(n+r-1)$, and let $\mathcal{O}_n := \{\mathbf{u}; \sum_{i=1}^{r+1} u_i v_i = 0, \text{ for all } \mathbf{v} \in \mathcal{L}_n\}$. Since $\dim \mathcal{L}_n \leq r$, it follows that $r+1 \geq \dim \mathcal{O}_n \geq 1$, for all $n \geq 0$. As a_0 is a unit, there exists an $M \in \mathbb{N}$ such that $a_0(n) \neq 0$ for all $n \geq M$. Let $j = \max\{n_0, M\}$. Then by (8.2.1), for all $n \geq j$,

$$y_i(n) = -\sum_{k=1}^{r} \frac{a_k(n)}{a_0(n)} y_i(n+k), \quad \text{for } i = 1, 2, \ldots, r+1.$$

Hence $\mathbf{y}(n)$ belongs to $\mathcal{L}_{n+1}$ when $n \geq j$. It follows that $\mathcal{L}_n \subseteq \mathcal{L}_{n+1}$ and $\mathcal{O}_{n+1} \subseteq \mathcal{O}_n$ for $n \geq j$, so that $\mathcal{O}_j \supseteq \mathcal{O}_{j+1} \supseteq \ldots$ is a decreasing chain of finite-dimensional linear subspaces. Every proper inclusion corresponds to a decrease in dimension; consequently there are only finitely many proper inclusions in the chain. Therefore there is an $m \in \mathbb{N}$ such that $\mathcal{O}_n = \mathcal{O}_m$ for all $n \geq m$. It follows that $\mathcal{O}_m$ is a subspace of $\mathcal{O}_n$ for every $n \geq j$. Since $\dim \mathcal{O}_n \geq 1$ for all $n \geq 0$, there is a nonzero vector $\mathbf{c} \in \mathcal{O}_m$. Thus $\mathbf{c} \in \cap_{n=j}^{\infty} \mathcal{O}_n$. This means that $y_1, y_2, \ldots, y_{r+1}$ are K-linearly dependent in $\mathcal{S}(K)$.

Now we show that $\dim \operatorname{Ker} L \geq r$. Since both a_r and a_0 are units of $\mathcal{S}(K)$, there exists an $n_0 \in \mathbb{N}$ such that $a_r(n), a_0(n) \neq 0$ for all $n \geq n_0$. Let $\mathbf{v}(0), \mathbf{v}(1), \ldots, \mathbf{v}(r-1)$ be a basis of K^r. Define sequences $y_1, y_2, \ldots, y_r \in \mathcal{S}(K)$ by

1. $y_i(n_0 + j) = v_i(j)$, $\qquad\qquad\qquad\qquad$ for $j = 0, \ldots, r-1$, $\quad$ (8.2.2)

2. $\qquad y_i(j) = -\sum_{k=0}^{r-1} \dfrac{a_k(j-r)}{a_r(j-r)} y_i(j-r+k)$, for $j \geq n_0 + r$, $\quad$ (8.2.3)

for $i = 1, 2, \ldots, r$. Multiplying (8.2.3) by $a_r(j-r)$ and setting $j - r = n$ shows that $Ly_i = 0$ for $i = 1, 2, \ldots, r$. We claim that $y_1, y_2, \ldots, y_r$ are linearly independent. Assume not. For all $n \geq 0$, let $\mathbf{y}(n) \in K^r$ be the vector with components

$y_1(n), y_2(n), \ldots, y_r(n)$. Denote by $\mathcal{L}_n$ the linear span of $\mathbf{y}(n), \mathbf{y}(n+1), \ldots, \mathbf{y}(n+r-1)$, and let $\mathcal{O}_n := \{\mathbf{u}; \sum_{i=1}^{r} u_i v_i = 0, \text{ for all } \mathbf{v} \in \mathcal{L}_n\}$. As in the preceding paragraph, we have $\mathcal{O}_{n+1} \subseteq \mathcal{O}_n$ for $n \geq n_0$. By our assumption of linear dependence there exists a nonzero vector $\mathbf{c} \in K^r$ such that $\mathbf{c} \in \mathcal{O}_n$ for all large enough n. It follows that $\mathbf{c} \in \mathcal{O}_n$ for all $n \geq n_0$. But by (8.2.2), $\mathbf{y}(n_0 + j) = \mathbf{v}(j)$ for $j = 0, 1, \ldots, r-1$, therefore $\mathcal{L}_{n_0} = K^r$ and $\mathcal{O}_{n_0} = \{0\}$, a contradiction. This proves the claim. $\qquad\square$

Definition 8.2.1 *A sequence $a \in \mathcal{S}(K)$ is* polynomial *over K if there is a polynomial $p(x) \in K[x]$ such that $a(n) = p(n)$ a.e. A sequence $a \in \mathcal{S}(K)$ is* rational *over K if there is a rational function $r(x) \in K(x)$ such that $a(n) = r(n)$ a.e. A nonzero sequence $a \in \mathcal{S}(K)$ is* hypergeometric *over K if there are nonzero polynomial sequences p and q over K such that $pNa + qa = 0$.*

We will denote the sets of polynomial, rational, and hypergeometric sequences over K by $\mathcal{P}(K), \mathcal{R}(K),$ and $\mathcal{H}(K)$, respectively. $\qquad\square$

Obviously, every polynomial sequence is rational, and every nonzero rational sequence is hypergeometric. Since a nonzero rational function has at most finitely many zeros, a nonzero rational sequence is always a unit.

Proposition 8.2.1 *Let $a, y \in \mathcal{S}(K)$, $y \neq 0$, and $Ny = ay$. Then both y and a are units.*

Proof. We have $y(n+1) = a(n)y(n)$ for all $n \geq n_0$, for some $n_0 \in \mathbb{N}$. If either $y_n = 0$ or $a_n = 0$ for some $n \geq n_0$, then $y(n) = 0$ for all $n \geq n_0 + 1$, thus $y = 0$, contrary to the assumption. It follows that both y and a are nonzero for all large enough n and hence are units. $\qquad\square$

Corollary 8.2.1 *Every hypergeometric sequence is a unit.*

Proof. Let $pNy + qy = 0$ where p and q are nonzero polynomials. By the remarks preceding Proposition 8.2.1, p is a unit. Therefore $Ny = -(q/p)y$ and $y \neq 0$. By Proposition 8.2.1, y is a unit. $\qquad\square$

8.3   Polynomial solutions

We wish to find all polynomial sequences y such that

$$Ly = f, \qquad\qquad (8.3.1)$$

where

$$L = \sum_{i=0}^{r} p_i(n) N^i \tag{8.3.2}$$

is a linear recurrence operator with polynomial coefficients $p_i \in \mathcal{P}(K)$, $p_r, p_0 \neq 0$, and f is a given sequence. How do we go about this?

First, if y is a polynomial then so is Ly. Therefore f had better be a polynomial sequence, or else we stand no chance. Second, we have already encountered a special case of this problem (with L of order 1) in Step 3 of Gosper's algorithm. Just as in that case, we split it into two subproblems:

1. Find an upper bound d for the possible degrees of polynomial solutions of (8.3.1).

2. Given d, describe all polynomial solutions of (8.3.1) having degree at most d.

To obtain a degree bound, it is convenient to rewrite L in terms of the difference operator $\Delta = N - 1$. Since $N = \Delta + 1$, we have

$$L = \sum_{i=0}^{r} p_i N^i = \sum_{i=0}^{r} p_i (\Delta + 1)^i = \sum_{i=0}^{r} p_i \sum_{j=0}^{i} \binom{i}{j} \Delta^j = \sum_{j=0}^{r} q_j \Delta^j,$$

where $q_j = \sum_{i=j}^{r} \binom{i}{j} p_i$. Let $y(n) = \sum_{k=0}^{d} a_k n^k$, where $a_k \in K$ and $a_d \neq 0$. Since[2] $\Delta^j n^k = k^{\underline{j}} n^{k-j} + O(n^{k-j-1})$, the leading coefficient of $q_j \Delta^j y(n)$ equals[3] $\mathrm{lc}\,(q_j) a_d d^{\underline{j}}$. Let

$$b := \max_{0 \leq j \leq r} (\deg q_j - j). \tag{8.3.3}$$

Clearly, $\deg Ly(n) \leq d + b$. If $d + b < 0$ then $d \leq -b - 1$ is the desired bound.[4] Otherwise the coefficient of n^{d+b} in $Ly(n)$ is

$$a_d \sum_{\substack{0 \leq j \leq r \\ \deg q_j - j = b}} \mathrm{lc}\,(q_j) d^{\underline{j}}.$$

We distinguish two cases: either $\deg Ly(n) = d + b$ and hence $d + b = \deg f$, or $\deg Ly(n) < d + b$ implying that the coefficient of n^{d+b} in $Ly(n)$ vanishes. This means that d is a root of the *degree polynomial*

$$\alpha(x) = \sum_{\substack{0 \leq j \leq r \\ \deg q_j - j = b}} \mathrm{lc}\,(q_j) x^{\underline{j}}. \tag{8.3.4}$$

---

[2] We use $a^{\underline{j}}$ for the *falling factorial* function $a(a-1)\ldots(a-j+1)$.

[3] $\mathrm{lc}\,(p)$ is the leading coefficient of the polynomial p.

[4] Note that b may be negative.

In each case, there is a finite choice of values that d can assume. In summary, we have

Proposition 8.3.1 *Let* $L = \sum_{i=0}^{r} p_i N^i$, $q_j = \sum_{i=j}^{r} \binom{i}{j} p_i$, *and suppose that* $Ly = f$, *where* f, y *are polynomials in* n. *Further let* $d_1 = \max\{x \in \mathbb{N}; \alpha(x) = 0\}$, *where* $\alpha(x)$ *is defined by (8.3.4). Then* $\deg y \leq d$, *where*

$$d = \max\{\deg f - b, -b - 1, d_1\}, \tag{8.3.5}$$

and b *is defined by (8.3.3).*

Once we have the degree bound d, the coefficients of polynomial solutions are easy to find: Set up a generic polynomial of degree d, plug it into the recurrence equation, equate the coefficients of like powers of n, and solve the resulting system of linear algebraic equations for $d+1$ unknown coefficients. This is called the *method of undetermined coefficients*.

Now we can state the algorithm.

Algorithm Poly

INPUT: Polynomials f and $p_i(n)$ over K, for $i = 0, 1, \ldots, r$.
OUTPUT: The general polynomial solution of (8.3.1) over K.

Step 1. Compute $q_j = \sum_{i=j}^{r} \binom{i}{j} p_i$, for $0 \leq j \leq r$.
Step 2. Compute d using (8.3.5).
Step 3. Using the method of undetermined coefficients, find all $y(n)$
      of the form $y(n) = \sum_{k=0}^{d} c_k n^k$ that satisfy (8.3.1).

Example 8.3.1. Let us find polynomial solutions of

$$3y(n+2) - ny(n+1) + (n-1)y(n) = 0. \tag{8.3.6}$$

Here $r = 2$ and $\deg f = -\infty$. In Step 1 we find that $q_0(n) = 2$, $q_1(n) = 6 - n$, and $q_2(n) = 3$. In Step 2 we compute $b = 0$ and $\alpha(x) = 2 - x$, hence $d = 2$. In Step 3 we obtain $C(n^2 - 11n + 27)$, where C is an arbitrary constant, as the general polynomial solution of (8.3.6). $\qquad\square$

Another, more sophisticated method that leads to a linear system with r unknowns is described in [ABP95]. When the degree d of polynomial solutions is large relative to the order r of the recurrence (as is often the case), this method may be considerably more efficient than the naïve one presented here.

8.4   Hypergeometric solutions

Let F be a field of characteristic zero and K an extension field of F. Given a linear recurrence operator L with polynomial coefficients over F, we seek solutions of

$$Ly = 0 \tag{8.4.1}$$

that are hypergeometric over K. We will call F the *coefficient field* of the recurrence. We assume that there exist algorithms for finding integer roots of polynomials over K and for factoring polynomials over K into factors irreducible over K.

Consider first the second-order recurrence

$$p(n)y(n+2) + q(n)y(n+1) + r(n)y(n) = 0. \tag{8.4.2}$$

Assume that $y(n)$ is a hypergeometric solution of (8.4.2). Then there is a rational sequence $S(n)$ such that $y(n+1) = S(n)y(n)$. Substituting this into (8.4.2) and cancelling $y(n)$ gives

$$p(n)S(n+1)S(n) + q(n)S(n) + r(n) = 0.$$

According to Theorem 5.3.1, we can write

$$S(n) = z \frac{a(n)}{b(n)} \frac{c(n+1)}{c(n)},$$

where $z \in K \setminus \{0\}$ and a, b, c are monic polynomials satisfying conditions (i), (ii), (iii) of that theorem. Then

$$z^2 p(n)a(n+1)a(n)c(n+2) + zq(n)b(n+1)a(n)c(n+1) + r(n)b(n+1)b(n)c(n) = 0. \tag{8.4.3}$$

The first two terms contain $a(n)$ as a factor, so $a(n)$ divides $r(n)b(n+1)b(n)c(n)$. By conditions (i) and (ii) of Theorem 5.3.1, $a(n)$ is relatively prime with $c(n)$, $b(n)$, and $b(n+1)$, so $a(n)$ divides $r(n)$. Similarly we find that $b(n+1)$ divides $p(n)a(n+1)a(n)c(n+2)$, therefore by conditions (i) and (ii) of Theorem 5.3.1, $b(n+1)$ divides $p(n)$. This leaves a finite set of candidates for $a(n)$ and $b(n)$: the monic factors of $r(n)$ and $p(n-1)$, respectively. We can cancel $a(n)b(n+1)$ from the coefficients of (8.4.3) to obtain

$$z^2 \frac{p(n)}{b(n+1)} a(n+1)c(n+2) + zq(n)c(n+1) + \frac{r(n)}{a(n)} b(n)c(n) = 0. \tag{8.4.4}$$

To determine the value of z, we consider the leading coefficient of the left hand side in (8.4.4) and find out that z satisfies a quadratic equation with known coefficients. So given the choice of $a(n)$ and $b(n)$, there are at most two choices for z.

For a fixed choice of $a(n)$, $b(n)$, and z, we can use algorithm Poly from page 154 to determine if (8.4.4) has any nonzero polynomial solution $c(n)$. If so, then we will have found a hypergeometric solution of (8.4.2). Checking all possible triples $(a(n), b(n), z)$ is therefore an algorithm which finds all hypergeometric solutions of (8.4.2). If the algorithm finds nothing, then this *proves* that (8.4.2) has no hypergeometric solution.

The algorithm that we have just derived for (8.4.2) easily generalizes to recurrences of arbitrary order.

Algorithm Hyper

INPUT: Polynomials $p_i(n)$ over F, for $i = 0, 1, \ldots, d$; an extension field K of F.

OUTPUT: A hypergeometric solution of (8.4.1) over K if one exists; 0 otherwise.

[1] For all monic factors $a(n)$ of $p_0(n)$ and $b(n)$ of $p_d(n - d + 1)$ over K do:

$P_i(n) := p_i(n) \prod_{j=0}^{i-1} a(n+j) \prod_{j=i}^{d-1} b(n+j)$, for $i = 0, 1, \ldots, d$;
$m := \max_{0 \le i \le d} \deg P_i(n)$;
let α_i be the coefficient of n^m in $P_i(n)$, for $i = 0, 1, \ldots, d$;

for all nonzero $z \in K$ such that

$$\sum_{i=0}^{d} \alpha_i z^i = 0 \qquad\qquad (8.4.5)$$

do:

    If the recurrence

$$\sum_{i=0}^{d} z^i P_i(n) c(n+i) = 0 \qquad\qquad (8.4.6)$$

    has a nonzero polynomial solution $c(n)$ over K then

        $S(n) := z(a(n)/b(n))(c(n+1)/c(n))$;
        return a nonzero solution $y(n)$ of $y(n+1) = S(n)y(n)$ and stop.

[2] Return 0 and stop. $\square$

Theorem 8.4.1 *Let $y(n)$ be a nonzero solution of (8.4.1) such that $y(n+1) = S(n)y(n)$ where $S(n)$ is a rational sequence. Let*

$$S(n) = z\frac{a(n)}{b(n)}\frac{c(n+1)}{c(n)},\qquad(8.4.7)$$

where a, b, c are monic polynomials satisfying conditions (i),(ii),(iii) of Theorem 5.3.1. Let $P_i(n)$ and α_i, for $i = 0, 1, \ldots, d$, be defined as in algorithm Hyper. Then

1. *$\sum_{i=0}^{d}\alpha_i z^i = 0$,*

2. *$a(n)$ divides $p_0(n)$,*

3. *$b(n)$ divides $p_d(n - d + 1)$, and*

4. *$c(n)$ satisfies (8.4.6).*

Proof. From (8.4.1) and $y(n+1) = S(n)y(n)$, it follows that

$$\sum_{i=0}^{d}p_i(n)\left(\prod_{j=0}^{i-1}S(n+j)\right)y(n) = 0,\qquad(8.4.8)$$

hence after cancelling $y(n)$ and using (8.4.7) we have

$$\sum_{i=0}^{d}p_i(n)z^i\left(\prod_{j=0}^{i-1}\frac{a(n+j)}{b(n+j)}\right)\frac{c(n+i)}{c(n)} = 0.\qquad(8.4.9)$$

Multiplication by $c(n)\prod_{j=0}^{d-1}b(n+j)$ now gives (8.4.6). All terms of the sum in (8.4.6) with $i > 0$ contain the factor $a(n)$, thus $a(n)$ divides the term with $i = 0$ which is $p_0(n)c(n)\prod_{j=0}^{d-1}b(n+j)$. By properties (i) and (ii) of the canonical form for rational functions (see Theorem 5.3.1 on page 84), it follows that $a(n)$ divides $p_0(n)$. Similarly, $b(n+d-1)$ divides $z^d p_d(n)c(n+d)\prod_{j=0}^{d-1}a(n+j)$, hence by properties (i) and (iii) of the same canonical form, $b(n+d-1)$ divides $p_d(n)$, so $b(n)$ divides $p_d(n-d+1)$. Finally, a look at the leading coefficient of the left hand side of (8.4.6) shows that $\sum_{i=0}^{d}\alpha_i z^i = 0$. $\qquad\square$

It is easy to see that the converse of Theorem 8.4.1 is also true, in the following sense: If z is an arbitrary constant, $a(n)$ and $b(n)$ are arbitrary sequences, $c(n)$ satisfies (8.4.6) where $P_i(n)$, for $i = 0, 1, \ldots, d$, is defined as in algorithm Hyper, and $y(n+1) = S(n)y(n)$ where $S(n)$ is as in (8.4.7), then $y(n)$ satisfies (8.4.1).

Example 8.4.1. In a recent Putnam competition, one of the problems was to find the general solution of

$$(n-1)y(n+2) - (n^2 + 3n - 2)y(n+1) + 2n(n+1)y(n) = 0.\qquad(8.4.10)$$

Let's try out Hyper on this recurrence. Here $p(n) = n - 1, q(n) = -(n^2 + 3n - 2), r(n) = 2n(n + 1)$. The monic factors of $r(n)$ are $1, n, n + 1$ and $n(n + 1)$, and those of $p(n - 1)$ are 1 and $n - 2$. Taking $a(n) = b(n) = 1$ yields $-z + 2 = 0$, hence $z = 2$. The auxiliary recurrence (8.4.4) is (after cancelling 2)

$$2(n - 1)c(n + 2) - (n^2 + 3n - 2)c(n + 1) + n(n + 1)c(n) = 0,$$

with polynomial solution $c(n) = 1$. This gives $S(n) = 2$ and $y(n) = 2^n$.

Taking $a(n) = n + 1$, $b(n) = 1$ yields $z^2 - z = 0$, hence $z = 1$ (recall that z must be nonzero). The auxiliary recurrence (8.4.4) is

$$(n - 1)(n + 2)c(n + 2) - (n^2 + 3n - 2)c(n + 1) + 2nc(n) = 0,$$

which again has polynomial solution $c(n) = 1$. This gives $S(n) = n + 1$ and $y(n) = n!$. We have found two linearly independent solutions of (8.4.10); we don't need to check the remaining possibilities for $a(n)$ and $b(n)$. Thus the general solution of (8.4.10) is

$$y(n) = C2^n + Dn!,$$

where C, D are arbitrary constants.   □

Alas, we are not always so lucky as in this example.

Example 8.4.2. In [vdPo79], it is shown[5] that the numbers

$$y(n) = \sum_{k=0}^{n} \binom{n}{k}^2 \binom{n + k}{k}^2 \qquad (8.4.11)$$

satisfy the recurrence

$$(n + 2)^3 y(n + 2) - (2n + 3)(17n^2 + 51n + 39)y(n + 1) + (n + 1)^3 y(n) = 0. \quad (8.4.12)$$

Here all the coefficients are of the same degree, therefore the equation for z will have no nonzero solution unless $a(n)$ and $b(n)$ are of the same degree as well. But they are both monic factors of $(n + 1)^3$, so they must be equal. Then the equation for z is $z^2 - 34z + 1 = 0$ with solutions $z = 17 \pm 12\sqrt{2}$. In either case, the auxiliary recurrence has no nonzero polynomial solutions, proving that (8.4.12) has no hypergeometric solution. As a consequence, (8.4.11) is not hypergeometric. □

When (8.4.1) has no hypergeometric solutions, we have to check all pairs of monic factors of the leading and trailing coefficient of L. The worst-case time complexity of Hyper is thus exponential in the degree of coefficients of (8.4.1). Nevertheless, a careful implementation can speed it up in several places. Here we give a few suggestions.

---

[5] Of course, Zeilberger's algorithm, of Chapter 6, will also find and prove this recurrence.

- We can reduce the degree of recurrence (8.4.6) by cancelling the factor $a(n)b(n+d-1)$, as we did in (8.4.4) for the case $d = 2$.

- Observe that the coefficients of equation (8.4.5) which determines z depend only on the difference $d(a,b) = \deg b(n) - \deg a(n)$ and not on $a(n)$ or $b(n)$ themselves. Therefore it is advantageous to test pairs of factors $a(n)$, $b(n)$ in order of the value of $d(a,b)$.

- We can skip those values of $d(a,b)$ for which (8.4.5) has a single nonzero term and thus no nonzero solution. For example, this happens when $\deg p_d(n) = \deg p_0(n) \geq \deg p_i(n)$ for $0 \leq i \leq d$, and $d(a,b) \neq 0$. Hence in this case it suffices to test pairs $a(n)$, $b(n)$ of equal degree (cf. Example 8.4.2).

- We can skip all pairs $a(n)$, $b(n)$ which do not satisfy property (i) of Theorem 5.3.1.

Example 8.4.3. The number $i(n)$ of involutions of a set with n elements satisfies the recurrence
$$y(n) = y(n-1) + (n-1)y(n-2).$$
More generally, let $r \geq 2$. The number $i_r(n)$ of permutations that contain no cycles longer than r satisfies the recurrence
$$y(n) = y(n-1) + (n-1)y(n-2) + (n-1)(n-2)y(n-3) + \ldots$$
$$+ (n-1)\cdots(n-r+2)(n-r+1)y(n-r). \qquad (8.4.13)$$

In Hyper, the degrees of the coefficients of auxiliary recurrences are obtained by adding to the degree sequence of the coefficients of the original recurrence (starting with the leading coefficient) an arithmetic progression with increment $D = d(a,b)$. In case of (8.4.13), the degree sequence is $0, 0, 1, 2, \ldots, r-1$. Adding to this sequence any arithmetic progression with integer increment D will produce a sequence with a single term of maximum value (the first one if $D < 0$; the last one if $D \geq 0$), implying that (8.4.5) has a single nonzero term for all choices of $a(n)$ and $b(n)$. Therefore (8.4.13) has no hypergeometric solution. This example shows that for any $d \geq 2$, there exist recurrences of order d without hypergeometric solutions. In particular, for $r = 2$, this means that the sum
$$i(n) = \sum_k \frac{n!}{(n-2k)!\, 2^k\, k!} \qquad (8.4.14)$$

(see, e.g., [Com74]), is not a hypergeometric term. $\qquad\qquad\square$

8.5   A Mathematica session

Algorithm Hyper is implemented in our Mathematica function Hyper[eqn, y[n]].
Here eqn is the equation and y[n] is the name of the unknown sequence. The
output from Hyper is a list of rational functions which represent the consecutive-
term ratios $y(n+1)/y(n)$ of hypergeometric solutions. qHyper is the q-analogue
of Hyper – it finds all q-hypergeometric solutions of q-difference equations with
rational coefficients (see [APP95]). It is available through the Web page for this
book (see Appendix A).

First we use Hyper on the Putnam recurrence (8.4.10).

```
In[9]:= Hyper[(n-1)y[n+2] - (n^2+3n-2)y[n+1] + 2n(n+1)y[n] == 0,
             y[n]]

Out[9]= {2}
```

This answer corresponds to $y(n) = 2^n$. But where is the other solution? We can
force Hyper to find all hypergeometric solutions by adding the optional argument
Solutions -> All.

```
In[10]:= Hyper[(n-1)y[n+2] - (n^2+3n-2)y[n+1] + 2n(n+1)y[n] == 0,
              y[n], Solutions -> All]

Out[10]= {2, 1 + n}
```

Now we can see the consecutive-term ratios of both hypergeometric solutions, 2^n
and $n!$. In general, Hyper[eqn, y[n], Solutions -> All] finds a generating set
(not necessarily linearly independent) for the space of closed form solutions of eqn.

Next, we return to Example 8.1.1 and use Hyper on the recurrence that we
found for $f(n)$ in Out[3].

```
In[11]:= Hyper[%3[[1]], SUM[n], Solutions -> All]
              27 (1 + n)    3 (2 + 3 n) (4 + 3 n)
Out[11]= {-----------, ---------------------}
              2 (3 + 2 n)    2 (1 + n) (3 + 2 n)
```

These two rational functions correspond to hypergeometric solutions $27^n/((2n+1)\binom{2n}{n})$ and $\binom{3n+1}{n}$. We check this for the latter solution:

```
In[12]:= FactorialSimplify[Binomial[3n+4,n+1]/Binomial[3n+1,n]]
              3 (2 + 3 n) (4 + 3 n)
Out[12]= ---------------------
              2 (1 + n) (3 + 2 n)
```

It follows that $f(n)$ is a linear combination of these two solutions. By comparing
the first two values, we determine that $f(n) = \binom{3n+1}{n}$, this time without advance
knowledge of the right hand side.

Now consider the following recurrence:

```
In[13]:= Hyper[y[n+2] - (2n+1)y[n+1] + (n^2-2)y[n] == 0, y[n]]
 Warning: irreducible factors of degree > 1 in trailing coefficient;
 some solutions may not be found
Out[13]= {}
```

Hyper found no hypergeometric solutions, but it printed out a warning that some
solutions may not have been found. In general, Hyper looks for hypergeometric
solutions over the rational number field $\mathbb{Q}$. However, by giving it the optional
argument Quadratics -> True we can force it to split quadratic factors in the
leading and trailing coefficients, and thus work over quadratic extensions of $\mathbb{Q}$.

```
In[14]:= Hyper[y[n+2]-(2n+1)y[n+1]+(n^2-2)y[n]==0, y[n],
        Quadratics->True, Solutions -> All]
Out[14]= {-Sqrt[2] + n, Sqrt[2] + n}
```

This means that there are two hypergeometric solutions[6] , $(\sqrt{2})_n$ and $(-\sqrt{2})_n$, over
$\mathbb{Q}(\sqrt{2})$.

8.6 Finding all hypergeometric solutions

Algorithm Hyper, as we stated it on page 156, stops as soon as it finds one hypergeometric solution. To find all solutions, we can check all possible triples $(a(n), b(n), z)$.
As it turns out, to obtain all hypergeometric solutions it suffices to take into account
only a basis of the space of polynomial solutions of the corresponding auxiliary recurrence for $c(n)$ (see Exercise 6).

   Another, better way to find all hypergeometric solutions is to find one with
Hyper, then reduce the order of the recurrence, recursively find solutions of the
reduced recurrence, and use Gosper's algorithm to put the antidifferences of these
solutions into closed form if possible. This method will actually yield a larger class
of solutions called *d'Alembertian sequences*. A sequence a is d'Alembertian if $a =
h_1 \sum h_2 \sum \cdots \sum h_k$ where $h_1, h_2, \ldots, h_k$ are hypergeometric terms, and $y = \sum x$
means that $\Delta y = x$. Alternatively, a sequence a is d'Alembertian if there are
first-order linear recurrence operators with rational coefficients $L_1, L_2, \ldots, L_k$ s.t.
$L_k L_{k-1} \cdots L_1 a = 0$ (see [AbP94]). It can be shown that d'Alembertian sequences
form a ring.

Example 8.6.1.   The number $d(n)$ of derangements (i.e., permutations without
fixed points) of a set with n elements satisfies the recurrence

$$y(n) = (n-1)y(n-1) + (n-1)y(n-2). \tag{8.6.1}$$

---

[6]We use $(a)_j$ for the *rising factorial* function $a(a+1)\ldots(a+j-1)$.

Taking $a(n) = b(n) = 1$ yields $z = -1$, but the auxiliary recurrence has no nonzero polynomial solution. The remaining choice $a(n) = n+1$, $b(n) = 1$ leads to $z^2 - z = 0$, so $z = 1$. The auxiliary recurrence

$$(n + 2)c(n + 2) - (n + 1)c(n + 1) - c(n) = 0$$

has, up to a constant factor, the only polynomial solution $c(n) = 1$. Then $S(n) = n + 1$ and

$$y(n) = n!$$

is, up to a constant factor, the only hypergeometric solution of (8.6.1). To reduce the order, we write $y(n) = z(n)n!$ where $z(n)$ is the new unknown sequence. Substituting this into (8.6.1) and writing $u(n) = z(n + 1) - z(n)$ yields

$$(n + 2)u(n + 1) + u(n) = 0,$$

a recurrence of order one. Taking $u(n) = (-1)^{n+1}/(n+1)!$ and $z(n) = \sum_{k=0}^{n}(-1)^k/k!$, we obtain another basic solution of (8.6.1) (which happens to be precisely the number of derangements):

$$d(n) = n! \sum_{k=0}^{n} \frac{(-1)^k}{k!} . \tag{8.6.2}$$

Now we apply Gosper's algorithm to the summand in (8.6.2) in order to put $d(n)$ into closed form. Since it fails, $d(n)$ is not a fixed sum of hypergeometric terms. Note, however, that $d(n)$ is a d'Alembertian sequence. $\qquad\square$

8.7   Finding all closed form solutions

Let $\mathcal{L}(\mathcal{H}(K))$ denote the K-linear hull of the set $\mathcal{H}(K)$ of all hypergeometric sequences.

Proposition 8.7.1 *Let L be as in (8.3.2), and let h be a hypergeometric term such that $Lh \neq 0$. Then Lh is hypergeometric and similar[7] to h.*

Proof. Let $r := Nh/h$. Then $N^i h = N^{i-1}(rh) = (N^{i-1}r)(N^{i-1}h) = \left(\prod_{j=0}^{i-1} N^j r\right) h$, so

$$Lh = \sum_{i=0}^{d} p_i N^i h = \left(\sum_{i=0}^{d} p_i \prod_{j=0}^{i-1} N^j r\right) h$$

is a nonzero rational multiple of h. $\qquad\square$

According to Proposition 5.6.3, every sequence from $\mathcal{L}(\mathcal{H}(K))$ can be written as a sum of pairwise dissimilar hypergeometric terms.

---

[7]See page 94.

Theorem 8.7.1 *Let L be a linear recurrence operator with polynomial coefficients, and $h \in \mathcal{L}(\mathcal{H}(K))$ such that $Lh = 0$. If $h = \sum_{i=1}^{k} h_i$ where h_i are pairwise dissimilar hypergeometric terms then*

$$Lh_i = 0, \quad for \quad i = 1, 2, \ldots, k.$$

Proof. By Proposition 8.7.1, for each i there exists a rational sequence r_i such that $Lh_i = r_i h_i$. Therefore

$$0 = Lh = \sum_{i=1}^{k} Lh_i = \sum_{i=1}^{k} r_i h_i.$$

Since the h_i are pairwise dissimilar, Theorem 5.6.1 implies that $r_i = 0$ for all i. $\square$

Corollary 8.7.1 *Let L be a linear recurrence operator with polynomial coefficients. Then the space $KerL \cap \mathcal{L}(\mathcal{H}(K))$ has a basis in $\mathcal{H}(K)$.*

Proof. Let $h \in \mathcal{L}(\mathcal{H}(K))$ satisfy $Lh = 0$. By Proposition 5.6.3, we can write $h = \sum_{i=1}^{k} h_i$ where h_i are pairwise dissimilar hypergeometric terms. By Theorem 8.7.1, each h_i satisfies $Lh_i = 0$. It follows that hypergeometric solutions of (8.4.1) span the space of solutions from $\mathcal{L}(\mathcal{H}(K))$. To obtain a basis for $KerL \cap \mathcal{L}(\mathcal{H}(K))$, select a maximal linearly independent set of hypergeometric solutions of (8.4.1). $\square$

From Theorem 8.4.1 and Corollary 8.7.1 it follows that the hypergeometric solutions returned by the recursive algorithm described in Section 8.6 constitute a basis for the space of solutions that belong to $\mathcal{L}(\mathcal{H}(K))$, i.e., *that algorithm finds all closed form solutions.*

We remark finally that if we are looking only for *rational* solutions of recurrences, then there is a more efficient algorithm for finding such solutions, due to S. A. Abramov [Abr95].

8.8 Some famous sequences that do not have closed form

Algorithm Hyper not only finds a spanning set for the space of closed form solutions, it also *proves*, if it returns the spanning set "∅", that a given recurrence with polynomial coefficients does not have a closed form solution. In this way we are able to prove that many well known combinatorial sequences cannot be expressed in closed form.

We must point out that the two notions of (a) having a closed form, as we have defined it (see page 145), and (b) having a pretty formula, do not quite coincide.

A good example of this is provided by the derangement function $d(n)$. This has no closed form, but it has the pretty formula $d(n) = \{n!/e\}$. This formula is not a hypergeometric term, and it is not a sum of a fixed number of same. But it sure is pretty!

The following theorem asserts that some famous sequences do not have closed forms. The reader will be able to find many more examples like these with the aid of programs `ct` and `Hyper`.

Theorem 8.8.1 *The following sequences cannot be expressed in closed form. That is to say, in each case the sequence cannot be exhibited as a sum of a fixed (independent of n) number of hypergeometric terms:*

- *The sum of the cubes of the binomial coefficients of order n, i.e., $\sum_k \binom{n}{k}^3$.*

- *The number of derangements (fixed-point free permutations) of n letters.*

- *The central trinomial coefficient, i.e., the coefficient of x^n in the expansion of $(1 + x + x^2)^n$.*

- *The number of involutions of n letters, i.e., the number of permutations of n letters whose square is the identity permutation.*

- *The sum of the "first third" of the binomial coefficients, i.e., $\sum_{k=0}^{n} \binom{3n}{k}$.*

First, for the sum $f(n) = \sum_k \binom{n}{k}^3$, program `ct` finds the recurrence

$$-8(n+1)^2 f(n) - (16 + 21n + 7n^2)f(n+1) + (n+2)^2 f(n+2) = 0,$$

(as well as a proof that this recurrence is correct, namely the two-variable recurrence for the summand). When we input this recurrence for $f(n)$ to Hyper, it returns the empty brackets "{}" that signify the absence of hypergeometric solutions.

The assertion as regards the central trinomial coefficients is left as an exercise (see Exercise 3) for the reader.

The fact that the number of involutions, $t(n)$, of n letters, is not of closed form is a special case of Example 8.4.3 on page 159.

The non-closed form nature of the number of derangements was shown in Example 8.6.1 on page 161.

The first third of the binomial coefficients, as well as many other possibilities, are left to the reader as easy exercises.   □

It is widely "felt" that for every $p \geq 3$ the sums of the pth powers of the binomial coefficients do not have closed form. The enterprising reader might wish to check this for some modest values of p. Many more possibilities for experimentation lie in the "pieces" of the full binomial sum

$$h(p,r) = \sum_{k=rn}^{(r+1)n} \binom{pn}{k},$$

whose status as regards closed form evaluation is unknown, in all of the non-obvious cases.

8.9 Inhomogeneous recurrences

In this section we show how to solve (8.3.1) over $\mathcal{L}(\mathcal{H}(K))$ when $f \neq 0$.

Proposition 8.9.1 *Up to the order of the terms, the representation of sequences from $\mathcal{L}(\mathcal{H}(K))$ as sums of pairwise dissimilar hypergeometric terms is unique.*

Proof. Assume that $a_1, a_2, \ldots, a_k$ and $b_1, b_2, \ldots, b_m$ are pairwise dissimilar hypergeometric terms with

$$\sum_{i=1}^{k} a_i = \sum_{j=1}^{m} b_j. \qquad (8.9.1)$$

Using induction on $k+m$ we prove that $k=m$ and that each a_i equals some b_j. If $k+m=0$ this holds trivially. Let $k+m>0$. Then by Theorem 5.6.1 it follows that $k>0$, $m>0$, and some a_i is similar to some b_j. Relabel the terms so that a_k is similar to b_m, and let $h := a_k - b_m$. If $h \neq 0$ then we can use induction hypothesis both on $\sum_{i=1}^{k-1} a_i + h = \sum_{j=1}^{m-1} b_j$ and $\sum_{i=1}^{k-1} a_i = \sum_{j=1}^{m-1} b_j - h$, to find that $k = m-1$ and $k-1 = m$. This contradiction shows that $h = 0$, so $a_k = b_m$ and $\sum_{i=1}^{k-1} a_i = \sum_{j=1}^{m-1} b_j$. By induction hypothesis, $k = m$ and each a_i with $1 \leq i \leq k-1$ equals some b_j with $1 \leq j \leq m-1$.   □

If $a \in \mathcal{L}(\mathcal{H}(K))$ and $La = f$ then $f \in \mathcal{L}(\mathcal{H}(K))$, by Proposition 8.7.1. Let $a = \sum_{j=1}^{m} a_j$ and $f = \sum_{j=1}^{k} f_j$ where a_j and f_j are pairwise dissimilar hypergeometric terms. Without loss of generality assume that there is an $l \leq m$ such that $La_j \neq 0$ if and only if $j \leq l$. Then by Proposition 8.9.1, $l = k$ and we can relabel the f_j so that

$$La_j = f_j, \quad \text{for } j = 1, 2, \ldots, k. \qquad (8.9.2)$$

By Proposition 8.7.1, there are nonzero rational sequences r_j such that $a_j = r_j f_j$, for $j = 1, 2, \ldots, k$. Let $s_j := Nf_j/f_j$. With L as in (8.3.2), it follows from (8.9.2) that r_j satisfies

$$L_j r_j = 1, \quad \text{for } j = 1, 2, \ldots, k \qquad (8.9.3)$$

where

$$L_j = \sum_{i=0}^{r} p_i \left(\prod_{l=0}^{i-1} N^l s_j \right) N^i, \quad \text{for} \quad j = 1, 2, \ldots, k.$$

This gives the following algorithm for solving (8.3.1) over $\mathcal{L}(\mathcal{H}(K))$:

1. Write $f = \sum_{j=1}^{k} f_j$ where f_j are pairwise dissimilar hypergeometric terms.

2. For $j = 1, 2, \ldots, k$, find a nonzero rational solution r_j of (8.9.3). If none exists for some j then (8.3.1) has no solution in $\mathcal{L}(\mathcal{H}(K))$.

3. Use Hyper to find a basis $a_1, a_2, \ldots, a_m$ for the space $\mathrm{Ker}L \cap \mathcal{L}(\mathcal{H}(K))$.

4. Return $\sum_{j=1}^{m} C_j a_j + \sum_{j=1}^{k} r_j f_j$ where C_j are arbitrary constants.

In Step 1 we need to group together similar hypergeometric terms, so we need to decide if a given hypergeometric term is rational. An algorithm for this is given by Theorem 5.6.2.

In Step 2 we use Abramov's algorithm mentioned on page 163. Note that from (8.9.2) it is easy to obtain *homogeneous* recurrences, at a cost of increasing the order by 1, satisfied by the a_j (see Exercise 10), which can then be solved by Hyper. However, using Abramov's algorithm in Step 2 is much more efficient.

8.10 Factorization of operators

Another application of algorithm Hyper is to the factorization of linear recurrence operators with rational coefficients, and construction of minimal such operators that annihilate definite hypergeometric sums. Recurrence operators can be multiplied using distributivity and the commutation rule

$$Np(n) = p(n+1)N.$$

Here $p(n)$ is considered to be an operator of order zero, and the apparent multiplication in the above equation is *operator* multiplication, rather than the application of operators to sequences.

To divide linear recurrence operators from the right, we use the formula

$$p(n)N^k = \left(\frac{p(n)}{q(n+k-m)} N^{k-m} \right) (q(n)N^m),$$

which follows immediately from the commutation rule. Here $p(n)$, $q(n)$ are rational functions of n, and $k \geq m$. Once we know how to divide monomials, operators can be divided as if they were ordinary polynomials in N. Consequently, for any two operators L_1, L_2 where $L_2 \neq 0$, there are operators Q and R such that $L_1 = QL_2 + R$ and $\operatorname{ord} R < \operatorname{ord} L_2$. Thus one can compute greatest common right divisors (and also least common left multiples, see [BrPe94]) of linear recurrence operators by the right-Euclidean algorithm.

If a sequence a is annihilated by some nonzero recurrence operator L with rational coefficients, then it is also annihilated by some nonzero recurrence operator M of minimal order and with rational coefficients. Right-dividing L by M we obtain operators Q and R such that $\operatorname{ord} R < \operatorname{ord} M$ and $L = QM + R$. Applying this to a we have $Ra = 0$. By the minimality of M, this is possible only if $R = 0$. Thus we have proved that the minimal operator of a right-divides any annihilating operator of a.

Solving equations is closely related to factorization of operators. Namely, if $Ly = 0$ then there is an operator L_2 such that $L = L_1 L_2$, where L_2 is the minimal operator for y. Conversely, if $L = L_1 L_2$ then any solution y of $L_2 y = 0$ satisfies $Ly = 0$ as well. In particular, if y is hypergeometric then its minimal operator is of order one and vice versa, hence we have a *one-to-one correspondence between hypergeometric solutions of $Ly = 0$ and monic first-order right factors of L*. For example, the two hypergeometric solutions 2^n and $n!$ of the Putnam recurrence (8.4.10) correspond to the two factorizations

$$(n-1)N^2 - (n^2 + 3n - 2)N + 2n(n+1) = ((n-1)N - n(n+1))(N-2)$$
$$= ((n-1)N - 2n)(N - (n+1)).$$

Furthermore, Hyper can be used to find first-order left factors of linear recurrence operators as well. If $L = \sum_{k=0}^d p_k(n) N^k$, then its *adjoint operator* is defined by

$$L^* = \sum_{k=0}^d p_{d-k}(n+k) N^k.$$

Simple computation shows that if L is as above then $L^{**} = N^d L N^{-d}$ and $(LM)^* = (N^d M^* N^{-d}) L^*$. Hence left factors of L correspond to right factors of L^*. More precisely, if $L^* = L_2 L_1$ where $\operatorname{ord} L_1 = 1$ then

$$L = N^{-d} L^{**} N^d$$
$$= N^{-d} (L_2 L_1)^* N^d$$
$$= N^{-d} (N^{d-1} L_1^* N^{1-d}) L_2^* N^d$$
$$= (N^{-1} L_1^* N)(N^{-d} L_2^* N^d).$$

Thus with Hyper we can find both right and left first-order factors. In particular, operators of orders 2 and 3 can be factored completely. As a consequence, we can find minimal operators for sequences annihilated by operators of orders 2 and 3.

Example 8.10.1. Let $\bar{a}_n$ denote the number of ways a random walk in the three-dimensional cubic lattice can return to the origin after $2n$ steps while always staying within $x \geq y \geq z$. In [WpZ89] it is shown that

$$\bar{a}_n = \sum_{k=0}^{n} \frac{(2n)! \, (2k)!}{(n-k)! \, (n+1-k)! \, k!^2 (k+1)!^2}.$$

Creative telescoping finds that $L\bar{a} = 0$ where

$$
\begin{aligned}
L = \; & 72(1+n)(2+n)(1+2n)(3+2n)(5+2n)(9+2n) \\
& - 4(2+n)(3+2n)(5+2n)(1377+1252n+381n^2+38n^3)N \\
& + 2(3+n)(4+n)^2(5+2n)(229+145n+22n^2)N^2 \\
& \qquad - (3+n)(4+n)^2(5+n)^2(7+2n)N^3, \quad (8.10.1)
\end{aligned}
$$

a recurrence operator of order 3. Hyper finds one hypergeometric solution of $Ly = 0$, namely

$$y(n) = \frac{(4n+7)(2n)!}{(n+1)! \, (n+2)!},$$

but looking at the first two values we see that $\bar{a}_n$ is not proportional to $y(n)$. Hence $\bar{a}_n$ is not hypergeometric, and its minimal operator has order 2 or 3.

Applying Hyper to L^* as described above we find that $L = L_1 L_2$ where

$$L_1 = 2(2+n)(9+2n) - (4+n)^2 N$$

and

$$
\begin{aligned}
L_2 = \; & 36(1+n)(1+2n)(3+2n)(5+2n) \\
& - 2(3+2n)(5+2n)(41+42n+10n^2)N \\
& \qquad + (2+n)(4+n)^2(5+2n)N^2. \quad (8.10.2)
\end{aligned}
$$

Note that $z = L_2\bar{a}$ satisfies the first-order equation $L_1 z = 0$. Since $z(0) = 0$ and the leading coefficient of L_1 does not vanish at nonnegative integers, it follows that $z = 0$. Thus (8.10.2) rather than (8.10.1) is a minimal operator annihilating the sequence $(\bar{a}_n)_{n=0}^{\infty}$.

$\square$

An algorithm for factorization of recurrence operators of any order is described in [BrPe94].

8.11   Exercises

1. In Example 8.1.1, replace the summand $F(n,k)$ by $(F(n,k) + F(n, n-k))/2$ in both cases. Note that the value of the sum does not change. Apply creative telescoping to the new summands. What are the orders of the resulting recurrences? (This *symmetrization trick* is essentially due to P. Paule [Paul94].)

2. Let $L_1 = (n-5)(n+1)N + (n-5)^2$, and $L_2 = (n-m)(n+1)N + (n-m)^2$. Find a basis of

   (a) $\operatorname{Ker} L_1$ in $K^{\mathbb{N}}$,

   (b) $\operatorname{Ker} L_1$ in $\mathcal{S}(\mathbb{Q})$,

   (c) $\operatorname{Ker} L_2$ in $\mathcal{S}(\mathbb{Q}(m))$.

3. Let $g(n)$ be the "central trinomial coefficient," i.e., the coefficient of x^n in the expansion of $(1 + x + x^2)^n$. Show that there is no simple formula for $g(n)$, as follows.

   (a) It is well known (see, e.g., Wilf [Wilf94], Ch. 5, Ex. 4) that

   $$g(n) = (\sqrt{3}/i)^n P_n(i/\sqrt{3}),$$

   where $P_n(x)$ is the nth Legendre polynomial. Use the formula for $P_n(x)$ given in Exercise 2 of Chapter 6 (page 121) to find a recurrence for the central trinomial coefficients $g(n)$.

   (b) Use algorithm Hyper to prove that there is no formula for the central trinomial coefficients that would express them as a sum of a fixed number of hypergeometric terms (this answers a question of Graham, Knuth and Patashnik [GKP89, 1st printing, Ch. 7, Ex. 56]).

4. In [GSY95] we encounter the sums

   $$f(n) = \sum_{k=0}^{n} \binom{3k}{k}\binom{3n-3k}{n-k},$$

   $$g(n) = \sum_{k=0}^{n-1} \binom{3k}{k}\binom{3n-3k-2}{n-k-1}.$$

   (a) Use creative telescoping to find recurrences satisfied by $f(n)$ and $g(n)$.

   (b) Use Hyper combined with reduction of order to express $f(n)$ and $g(n)$ in terms of sums in which the running index n does not appear under the summation sign.

(c) What is the minimum order of a linear operator with polynomial coefficients annihilating $g(n)$?

5. Solve $n(n+1)y(n+2) - 2n(n+k+1)y(n+1) + (n+k)(n+k+1)y(n) = 0$ over the field $\mathbb{Q}(k)$ where k is transcendental over $\mathbb{Q}$.

6. Solve $a(n+2) - (2n+1)a(n+1) + (n^2 - u)a(n) = 0$ over $\mathbb{Q}(\sqrt{u})$.

7. Let r, h, h_i, c, c_i be nonzero sequences from $S(K)$ such that $\frac{h(n+1)}{h(n)} = r(n)\frac{c(n+1)}{c(n)}$, $\frac{h_i(n+1)}{h_i(n)} = r(n)\frac{c_i(n+1)}{c_i(n)}$, for $i = 1, 2, \ldots, k$, and c is a K-linear combination of c_i. Show that h is a K-linear combination of h_i.

8. Prove that the sequence of Fibonacci numbers defined by $f(0) = f(1) = 1$, $f(n+2) = f(n+1) + f(n)$ is not hypergeometric over any field of characteristic zero.

9. Show that by a suitable hypergeometric substitution, any linear recurrence with polynomial coefficients can be turned into one with unit leading coefficient.

10. Let L be a linear difference operator of order d with rational coefficients over K. Let y be a sequence from $S(K)$ such that $Ly = f$ is hypergeometric.

    (a) Find an operator M of order $d+1$ such that $My = 0$.

    (b) If $\{a_1, a_2, \ldots, a_d\}$ is a basis for the kernel of L, find a basis for the kernel of M.

Solutions

1. 1 and 2, respectively.

2. (a) $\{a_1, a_2\}$ where $a_1(n) = \binom{5}{n}$, $a_2(n) = \begin{cases} 0, & \text{for } n < 6 \\ (-1)^n/\binom{n}{6}, & \text{for } n \geq 6 \end{cases}$

   (b) $\{a\}$ where $a(n) = (-1)^n/\binom{n}{6}$, for $n \geq 6$

   (c) $\{a\}$ where $a(n) = \binom{m}{n}$

4. (a) $Lf = Mg = 0$ where

$$L = 8(n+2)(2n+3)N^2 - 6(36n^2 + 99n + 70)N + 81(3n+2)(3n+4),$$
$$M = 16(n+2)(2n+3)(2n+5)N^3 - 12(2n+3)(54n^2 + 153n + 130)N^2$$
$$+ 324(27n^3 + 72n^2 + 76n + 30)N - 2187n(3n+1)(3n+2).$$

(b) For $Ly = 0$ Hyper finds one solution $(27/4)^n$. After reducing the order and matching initial conditions we have

$$f(n) = \frac{1}{2} \left(\frac{27}{4}\right)^n \left(1 - \sum_{k=0}^{n} \frac{\binom{3k}{k}}{3k-1} \left(\frac{27}{4}\right)^{-k}\right), \quad \text{for } n \geq 0.$$

For $My = 0$ Hyper finds two solutions, $(27/4)^n$ and $27^n/(n\binom{2n}{n})$. After reducing the order and matching initial conditions we have

$$g(n) = \frac{1}{27} \left(\frac{27}{4}\right)^n \left(3 + \sum_{k=0}^{n-1} \frac{\binom{3k}{k}}{2k+1} \left(\frac{27}{4}\right)^{-k}\right), \quad \text{for } n \geq 1.$$

(c) Since $g(n)$ does not belong to the linear span of the two hypergeometric solutions of $My = 0$, the order of a minimal annihilator is either 2 or 3. Using the above expression for $g(n)$ we determine that

$$M_1 g(n) = \frac{1}{2n+1} \binom{3n}{n}$$

where $M_1 = 4N - 27$, hence $g(n)$ is annihilated by the second-order operator $M_2 M_1$ where $M_2 = 2(n+1)(2n+3)N - 3(3n+1)(3n+2)$ annihilates $\binom{3n}{n}/(2n+1)$.

5. $y(n) = C_1 \binom{n+k-1}{n-1} + C_2 \binom{n+k-1}{n-2}$

6. $a(n) = C_1(+\sqrt{u})_n + C_2(-\sqrt{u})_n$

9. Let $\sum_{k=0}^{d} p_k(n)y(n+k) = f(n)$ where $p_i(n)$ are polynomials. If

$$y(n) = \frac{x(n)}{\prod_{j=j_0}^{n-d} p_d(j)}$$

then $x(n+d) + \sum_{k=0}^{d-1} p_k(n) \left(\prod_{j=1}^{d-k-1} p_d(n-j)\right) x(n+k) = f(n) \prod_{j=j_0}^{n-1} p_d(j).$

10. (a) Let L_1 be a first-order operator such that $L_1(f) = 0$. Take $M = L_1 L$.

   (b) $\{a_1, a_2, \ldots, a_d, y\}$

Part III

Epilogue

Chapter 9

An Operator Algebra Viewpoint

9.1 Early history

Quite early people recognized that, say, four sticks are more than three sticks, and likewise, four stones are more than three stones. Only much later was it noticed that these two inequalities are "isomorphic," and that a collection of three stones has something in common with a collection of three sticks, viz. "threeness." Thus was born the very abstract notion of *number*.

Then came problems about numbers. "My age today is four times the age of my daughter. In twenty years, it would be only twice as much." It was found that rather than keep *guessing and checking*, until hitting on the answer, it is useful to call the yet unknown age of the daughter by a *symbol*, x, set up the equation: $4x + 20 = 2(x + 20)$, and *solve for x*. Thus algebra was born. Expressions in the symbol x that used only addition, subtraction and multiplication were called *polynomials* and soon it was realized that one can add and multiply (but not, in general, divide) polynomials, just as we do with numbers.

Then came problems about *several* (unknown) numbers, which were usually denoted by x, y, z. After setting up the equations, one got a *system* of equations, like

$$\text{(i) } 2x + y + z = 6 \quad \text{(ii) } x + 2y + z = 5 \quad \text{(iii) } x + y + 2z = 5. \qquad (9.1.1)$$

The subject that treats such equations, in which all the unknowns occur *linearly*, is called *linear algebra*. Its central idea is to unite the separate unknown quantities into one entity, the *vector*, and to define an operation that takes vectors into vectors that mimics multiplication by a fixed number. This led to the revolutionary concept of *matrix* (due to Cayley and Sylvester). In linear algebra, (9.1.1) is shorthanded to

$$\begin{pmatrix} 2 & 1 & 1 \\ 1 & 2 & 1 \\ 1 & 1 & 2 \end{pmatrix} \begin{pmatrix} x \\ y \\ z \end{pmatrix} = \begin{pmatrix} 6 \\ 5 \\ 5 \end{pmatrix}.$$

Alternatively, we can find x by *eliminating* y and z. First we eliminate y, to get

$$(\mathrm{i}') := 2(\mathrm{i}) - (\mathrm{ii}) := 3x + z = 7 \quad (\mathrm{ii}') := (\mathrm{ii}) - 2(\mathrm{iii}) := -x - 3z = -5. \quad (9.1.2)$$

We next eliminate z:

$$(\mathrm{iii}') := 3(\mathrm{i}') + (\mathrm{ii}') := 8x = 16,$$

from which it follows immediately that $x = 2$.

Similarly, we can find y and z. Once found, it is trivial to verify that $x = 2, y = 1, z = 1$ indeed satisfies the system (9.1.1), but to find that solution required ingenuity, or so it seemed.

Then it was realized, probably before Gauss, that one can do this systematically, by performing *Gaussian elimination*, and it all became routine.

What if you have several unknowns, say, x, y, z, w, and you have a system of *non-linear* equations? If the equations are *polynomial*, say,

$$P(x, y, z, w) = 0, \quad Q(x, y, z, w) = 0, \quad R(x, y, z, w) = 0, \quad S(x, y, z, w) = 0.$$

Then it is still possible to perform elimination. Sylvester gave such an algorithm, but a much better, beautiful, algorithm was given by Bruno Buchberger [Buch76], the celebrated *Gröbner Basis* algorithm.

9.2   Linear difference operators

Matrices induce *linear operators* that act on *vectors*. What is a vector? An n-component vector is a function from the finite set $\{1, 2, \ldots, n\}$ into the set of numbers (or, more professionally, into a field).

When we replace a finite dimensional vector by an infinite one, we get a *sequence*, which is a function defined on the natural numbers $\mathbf{N}$, or more generally, on the integers. Traditionally sequences were denoted by using subscripts, like a_n, unlike their continuous counterparts $f(x)$, in which the argument was at the same level. Being proponents of discrete-lib, we may henceforth write $a(n)$ for a sequence. For example, the Fibonacci numbers will be denoted by $F(n)$ rather than F_n.

Recall that a linear operator induced by a matrix A is an operation that takes a vector $x(i), i = 1....., n$, and sends it to a vector $y(i), i = 1....., n$, given by

$$y(i) = \sum_{j=1}^{n} a_{i,j} x(j) \quad (i = 1, \ldots, n).$$

Thus, in general, each and every $x(j)$ influences the value of each $y(i)$. A *linear recurrence operator* is the analog of this for *infinite* sequences in which the value of $y(n)$ depends only on those $x(m)$ for which m is not too far from n. In other words, it has the form

$$y(n) := \sum_{j=-L}^{M} a(n,j) x(n+j), \tag{9.2.1}$$

where L and M are pre-determined nonnegative integers.

The simplest linear recurrence operator, after the identity and zero operators, is the one that sends $x(n)$ to $x(n+1)$: The value of y today is the value of x tomorrow. We will denote it by E_n, or N. Thus

$$Nx(n) := x(n+1) , \quad (n \in \mathbb{Z}).$$

Its inverse is the yesterday operator

$$N^{-1}x(n) := x(n-1) \quad (n \in \mathbb{Z}).$$

Iterating, we get that for *every* (positive, negative, or zero) integer,

$$N^r x(n) := x(n+r) \quad (n \in \mathbb{Z}).$$

In terms of this *fundamental shift operator* N, the general linear recurrence operator in (9.2.1), $x(n) \to y(n)$, let's call it A, can be written

$$A := \sum_{j=-L}^{M} a(n,j) N^j. \tag{9.2.2}$$

Linear operators in linear algebra can be represented by matrices A, where the operation is $x(n) \to Ax(n)$. It proves convenient to talk about A both *qua* matrix and *qua* linear operator, *without* mentioning the vector $x(n)$ that it acts on. Then we can talk about matrix algebra, and multiply matrices *per se*. We are also familiar with this abstraction process from calculus, where we sometimes write f for a function, without committing ourselves to naming the argument, as in $f(x)$. Likewise, the operation of differentiation is denoted by D, and we write $D(\sin) = \cos$.

Since the range of j is finite, it is more convenient to rewrite (9.2.2) as

$$A := \sum_{j=-L}^{M} a_j(n)N^j. \qquad (9.2.3)$$

So a *linear recurrence operator* is just a *Laurent polynomial* in N, with coefficients that are discrete functions of n. The class of all such operators is a non-commutative algebra, where the addition is the obvious one and multiplication is performed on monomials by $(a(n)N^r)(b(n)N^s) = a(n)b(n+r)N^{r+s}$, and extended linearly. So if

$$B := \sum_{j=-L}^{M} b_j(n)N^j,$$

then

$$AB := \sum_{k=-2L}^{2M} \left(\sum_{j=-L}^{M} a_j(n)b_{k-j}(n+j) \right) N^k.$$

For example

$$(1 + e^n N)(1 + |n|N) = 1 + (e^n + |n|)N + (e^n|n+1|)N^2.$$

As with matrices, operator notation started out as shorthand, but then turned out to be much more. We have seen and will soon see again some non-trivial applications, but for now let's have a trivial one.

Example 9.2.1. Prove that the Fibonacci numbers $F(n)$ satisfy the recurrence

$$F(n+4) = F(n+2) + 2F(n+1) + F(n).$$

Verbose Proof.

$$\begin{aligned}
&\text{(i)} \quad F(n+2) - F(n+1) - F(n) \quad\ \ = 0, \\
&\text{(ii)} \quad F(n+3) - F(n+2) - F(n+1) = 0, \\
&\text{(iii)} \quad F(n+4) - F(n+3) - F(n+2) = 0.
\end{aligned}$$

Adding (i), (ii), (iii), we get

$$\text{(i)} + \text{(ii)} + \text{(iii)} : F(n+4) - F(n+2) - 2F(n+1) - F(n) = 0.$$

Terse Proof.

$$\begin{aligned}
(N^2 - N - 1)F(n) = 0 \quad &\Rightarrow \quad (N^2 + N + 1)(N^2 - N - 1)F(n) = 0 \\
&\Rightarrow \quad (N^4 - N^2 - 2N - 1)F(n) = 0. \qquad \square
\end{aligned}$$

In linear algebra, the primary objects are vectors, and matrices only help in making sense of personalities and social lives; in this kind of algebra, the primary objects are not operators, but *sequences*, and the relations between them.

Given a linear recurrence operator

$$A = \sum_{j=0}^{L} a_j(n) N^j,$$

we are interested in sequences $x(n)$ that are *annihilated* by A, i.e., sequences for which $Ax(n) \equiv 0$. In longhand, this means

$$\sum_{j=0}^{L} a_j(n) x(n+j) = 0 \quad (n \geq 0).$$

Once a sequence, $x(n)$, is a solution of one linear recurrence equation, $Ax = 0$, it is a solution of infinitely many equations, namely $BAx = 0$, for every linear recurrence operator B.

Notice that the collection of all linear recurrence operators is a ring, and the above remark says that, for any sequence, the set of operators that annihilate it is (a possibly trivial) *ideal.*

Alas, every sequence is annihilated by *some* operator: If $x(n)$ is an arbitrary sequence none of whose terms vanish, then, tautologically, $x(n)$ is annihilated by the linear difference operator $N - (x(n+1)/x(n))$, which is first-order, to boot! In order to have an interesting theory of sequences, we have to be more exclusive, and proclaim that a sequence $x(n)$ is *interesting* if it is annihilated by a linear difference operator with *polynomial* coefficients. From now on, until further notice, all the coefficients of our recurrence operators will be *polynomials.*

There is a special name for such sequences. In fact there are two names. The first name is *P-recursive*, and the second name is *holonomic.* The reason for the first name (coined, we believe, by Richard Stanley [Stan80]) is clear, the "P-" standing for "Polynomial". The term "holonomic," coined in [Zeil90a], is by analogy with the theory of holonomic differential equations ([Bjor79, Cart91]). Meanwhile, let us make it an official definition.

Definition 9.2.1 *A sequence $x(n)$ is* P-recursive, *or* holonomic, *if it is annihilated by a linear recurrence operator with polynomial coefficients. In other words, if there exist a nonnegative integer L, and polynomials $p_0(n), \ldots, p_L(n)$ such that*

$$\sum_{i=0}^{L} p_i(n) x(n+i) = 0 \quad (n \geq 0).$$

Many sequences that arise in combinatorics happen to be P-recursive. It is useful to be able to "guess" the recurrence empirically. There is a program to do this in the Maple package gfun written by Bruno Salvy and Paul Zimmerman. That package is in the Maple Share library that comes with Maple V, versions 3 and up. Another version can be found in the program findrec in the Maple package EKHAD that comes with this book. The function call is

$$\text{findrec(f,DEGREE,ORDER,n,N)}$$

where f is the beginning of a sequence, written as a list, DEGREE is the maximal degree of the coefficients, ORDER is the guessed order of the recurrence, n is the symbol denoting the subscript (variable), and N denotes the shift operator in n. The last two arguments are optional. The defaults are the symbols n and N.

For example

$$\text{findrec}([1,1,1,1,1,1,1,1,1],0,1)$$

yields the output $-1 + N$, while

$$\text{findrec}([1,1,2,3,5,8,13,21,34],0,2)$$

yields the recurrence $-1 - N + N^2$, and

$$\text{findrec}([1,2,6,24,120,720,5040,8!,9!],1,1);$$

yields $(1+n) - N$.

Exercise. Use findrec to find, empirically, recurrences satisfied by the "log 2" sequence

$$\sum_{k=0}^{n} \binom{n}{k}\binom{n+k}{k},$$

with DEGREE= 1 and ORDER= 2; by Apéry's "$\zeta(2)$" sequence

$$\sum_{k=0}^{n} \binom{n}{k}^2 \binom{n+k}{k},$$

with DEGREE= 2 and ORDER= 2; and by Apéry's "$\zeta(3)$" sequence

$$\sum_{k=0}^{n} \binom{n}{k}^2 \binom{n+k}{k}^2,$$

with DEGREE= 3 and ORDER= 2.

The sequences $\{2^n\}$ and the Fibonacci numbers $\{F(n)\}$ are obviously P-recursive, in fact they are C-recursive, because the coefficients in their recurrences are not only polynomials, they are constants. Other obvious examples are $\{n!\}$ and the Catalan numbers $\{\frac{1}{n+1}\binom{2n}{n}\}$, in which the relevant recurrence is *first order*, i.e., $L = 1$. There is a special name for such distinguished sequences: *hypergeometric*. Note that the following definition is just a rephrasing of our earlier definition of the same concept on page 34.

Definition 9.2.2 *A sequence $x(n)$ is called* hypergeometric *if it is annihilated by a first-order linear recurrence operator with polynomial coefficients, i.e., if there exist polynomials $p_0(n)$, and $p_1(n)$ such that*

$$p_0(n)x(n) + p_1(n)x(n+1) = 0 \qquad (n \geq 0).$$

Yet another way of saying the same thing is that a sequence $x(n)$ is hypergeometric if $x(n+1)/x(n)$ is a rational function of n. This explains the reason for the name *hypergeometric* (coined, we believe, by Gauss). A sequence $x(n)$ is called *geometric* if $x(n+1)/x(n)$ is a *constant*, and allowing rational functions brings in the hype.

An example of a P-recursive sequence that is *not* hypergeometric is the number of permutations on n letters that are involutions, i.e., that consist of 1- and 2-cycles only. This sequence, $t(n)$, obviously satisfies

$$t(n) = t(n-1) + (n-1)t(n-2),$$

as one sees by considering separately those n-involutions in which the letter n is a fixed point (a 1-cycle), and those in which n lives in a 2-cycle.

The reader of this book knows by now that in addition to sequences of one discrete variable $x(n)$, like 2^n and $F(n)$, we are interested in multivariate sequences, like $\binom{n}{k}$. A multi-sequence $F(n_1, \ldots, n_k)$, of k discrete variables, is a function on k-tuples of integers. Depending on the context, the n_i will be nonnegative or arbitrary integers. A famous example is the multisequence of *multinomial coefficients*:

$$\binom{n_1 + \cdots + n_k}{n_1, \ldots, n_k} := \frac{(n_1 + \cdots + n_k)!}{n_1! \ldots n_k!}.$$

We propose now to discuss the important subject of elimination, in order to explain how it can be used to find recurrence relations for sums. Before discussing elimination in the context of arbitrarily many variables $F(n_1, \ldots, n_k)$, we will cover, in some detail, the very important case of two variables.

9.3 Elimination in two variables

Let's take the two variables to be (n, k). The shift operators N, K act on discrete functions $F(n, k)$, by

$$NF(n, k) := F(n + 1, k); \qquad KF(n, k) := F(n, k + 1).$$

For example, the Pascal triangle equality

$$\binom{n + 1}{k + 1} = \binom{n}{k + 1} + \binom{n}{k}$$

can be written, in operator notation, as

$$(NK - K - 1)\binom{n}{k} = 0.$$

If a discrete function $F(n, k)$ satisfies two partial linear recurrences

$$P(N, K, n, k)F(n, k) = 0, \quad Q(N, K, n, k)F(n, k) = 0,$$

then it satisfies many, many others:

$$\{A(N, K, n, k)P(N, K, n, k) + B(N, K, n, k)Q(N, K, n, k)\}F(n, k) = 0, \quad (9.3.1)$$

where A and B can be any linear partial recurrence operators.

So far, everything has been true for arbitrary linear recurrence operators. From now on we will only allow linear recurrence operators *with polynomial coefficients*. The set $C\langle n, k, N, K \rangle$ of all linear recurrence operators with polynomial coefficients is a non-commutative, associative algebra generated by N, K, n, k subject to the relations

$$NK = KN, \qquad Nk = kN, \qquad nK = Kn, \qquad (9.3.2)$$

$$nk = kn, \qquad Nn = (n + 1)N, \qquad Kk = (k + 1)K. \qquad (9.3.3)$$

Under certain technical conditions on the operators P and Q (viz. *holonomicity* [Zeil90a, Cart91]) we can, by a clever choice of operators A and B, get the operator in the braces in (9.3.1), call it $R(N, K, n)$, to be independent of k. This is called *elimination*.

Now write

$$R(N, K, n) = S(N, n) + (K - 1)\bar{R}(N, K, n)$$

(where $S(N, n) := R(N, 1, n)$). Since $R(N, K, n)F(n, k) \equiv 0$, we have

$$S(N, n)F(n, k) = (K - 1)[-\bar{R}(N, K, n)F(n, k)].$$

If we call the function inside the above square brackets $G(n, k)$, we get

$$S(N, n)F(n, k) = (K - 1)G(n, k).$$

If $F(n, \pm\infty) = 0$ for every n and the same is true of $G(n, \pm\infty)$, then summing the above w.r.t. k yields

$$S(N, n)(\sum_k F(n, k)) - \sum_k (G(n, k+1) - G(n, k)) = 0.$$

So

$$a(n) := \sum_k F(n, k)$$

satisfies the recurrence

$$S(N, n)a(n) = 0.$$

Example 9.3.1.

$$F(n, k) = \frac{n!}{k! \, (n - k)!}$$

First let us find operators P and Q that annihilate F. Since

$$\frac{F(n + 1, k)}{F(n, k)} = \frac{n + 1}{n - k + 1} \quad \text{and}$$
$$\frac{F(n, k + 1)}{F(n, k)} = \frac{n - k}{k + 1},$$

we have

$$(n - k + 1)F(n + 1, k) - (n + 1)F(n, k) = 0,$$
$$(k + 1)F(n, k + 1) - (n - k)F(n, k) = 0.$$

In operator notation,

(i) $[(n - k + 1)N - (n + 1)]F \equiv 0$, (ii) $[(k + 1)K - (n - k)]F \equiv 0$.

Expressing the operators in descending powers of k, we get

(i) $[(-N)k + (n + 1)N - (n + 1)]F \equiv 0$, (ii) $[(K + 1)k - n]F \equiv 0$.

Eliminating k, we obtain

$$(K + 1)(\mathrm{i}) + N(\mathrm{ii}) = \{(K + 1)[(n + 1)N - (n + 1)] + N(-n)\}F \equiv 0,$$

which becomes

$$(n+1)[NK - K - 1]F \equiv 0.$$

We have that

$$R(N,K,n) = (n+1)[NK - K - 1]; \quad S(N,n) = R(N,1,n) = (n+1)[N - 2],$$

and therefore we have proved the deep result that

$$a(n) := \sum_k \binom{n}{k}$$

satisfies

$$(n+1)(N - 2)a(n) \equiv 0,$$

i.e., in everyday notation, $(n+1)[a(n+1) - 2a(n)] \equiv 0$, and hence, since $a(0) = 1$, we get that $a(n) = 2^n$.   □

Important observation of Gert Almkvist. So far we have had two stages:

$$R(N,K,n) = A(N,K,n,k)P(N,K,n,k) + B(N,K,n,k)Q(N,K,n,k)$$
$$R(N,K,n) = S(N,n) + (K-1)\bar{R}(N,K,n),$$

i.e.,

$$S(N,n) = AP + BQ + (K - 1)(-\bar{R}),$$

where $\bar{R}$ has the nice but *superfluous* property of not involving k. WHAT A WASTE! So we are led to formulate the following.

9.4   Modified elimination problem

Input: Linear partial recurrence operators with polynomial coefficients $P(N,K,n,k)$ and $Q(N,K,n,k)$. Find operators A, B, C such that

$$S(N,n) := AP + BQ + (K - 1)C$$

does not involve K and k.

Remark. Note something strange: We are allowed to multiply P and Q by any operator *from the left*, but not from the right, while we are allowed to multiply $K - 1$ by any operator *from the right*, but not from the left. In other words, we have to find a non-zero operator, depending on n and N only, in the ambidextrous "ideal" generated by $P, Q, K - 1$, but of course this is not an ideal at all. It would be very nice if one had a Gröbner basis algorithm for doing that. Nobuki Takayama made considerable progress in [Taka92].

Let a discrete function $F(n,k)$ be annihilated by two operators P and Q that are "independent" in some technical sense (i.e., they form a holonomic ideal, see [Zeil90a, Cart91]). Performing the elimination process above (and the holonomicity guarantees that we'll be successful), we get the operators A, B, C and $S(N,n)$. Now let

$$G(n,k) = C(N,K,n,k)F(n,k).$$

We have

$$S(N,n)F(n,k) = (K-1)G(n,k).$$

It follows that

$$a(n) := \sum_k F(n,k)$$

satisfies

$$S(N,n)a(n) \equiv 0.$$

Let's apply the elimination method to find a recurrence operator annihilating $a(n)$, with

$$F(n,k) := \binom{n}{k}\binom{b}{k} = \frac{n!\,b!}{k!^2(n-k)!\,(b-k)!},$$

and thereby prove and discover the Chu–Vandermonde identity.

We have

$$\frac{F(n+1,k)}{F(n,k)} = \frac{(n+1)}{(n-k+1)},$$
$$\frac{F(n,k+1)}{F(n,k)} = \frac{(n-k)(b-k)}{(k+1)^2}.$$

Cross multiplying,

$$(n-k+1)F(n+1,k) - (n+1)F(n,k) = 0,$$
$$(k+1)^2F(n,k+1) - (n-k)(b-k)F(n,k) = 0.$$

In operator notation:

$$((n-k+1)N - (n+1))F = 0$$
$$((k+1)^2K - (nb-bk-nk+k^2))F = 0.$$

So F is annihilated by the two operators P and Q, where

$$P = (n-k+1)N - (n+1); \quad Q = (k+1)^2K - (nb-bk-nk+k^2).$$

We would like to find a good operator that annihilates F. By *good* we mean "independent of k," modulo $(K - 1)$ (where the multiples of $(K - 1)$ that we are allowed to throw out are right multiples).

Let's first write P and Q in ascending powers of k:

$$P = (-N)k + (n + 1)N - (n + 1)$$
$$Q = (n + b)k - nb + (K - 1)k^2,$$

and then eliminate k modulo $(K - 1)$. However, we must be careful to remember that right multiplying a general operator G by $(K - 1)$STUFF does not yield, in general, $(K - 1)$STUFF$'$. In other words,

Warning:

$$\text{OPERATOR}(N, K, n, k)(K - 1)(\text{STUFF}) \neq (K - 1)(\text{STUFF}').$$

Left multiplying P by $n + b + 1$, left multiplying Q by N and adding yields

$$(n + b + 1)P + NQ = (n + b + 1)[-Nk + (n + 1)N - (n + 1)]$$
$$+ N[(n + b)k - nb + (K - 1)k^2]$$
$$= (n + 1)[(n + 1)N - (n + b + 1)] + (K - 1)[Nk^2].$$

So, in the above notation,

$$S(N, n) = (n + 1)[(n + 1)N - (n + b + 1)], \ \bar{R} = Nk^2. \tag{9.4.1}$$

It follows that

$$a(n) := \sum_k \binom{n}{k}\binom{b}{k}$$

satisfies

$$((n + 1)N - (n + b + 1))a(n) \equiv 0,$$

or, in everyday notation,

$$(n + 1)a(n + 1) - (n + b + 1)a(n) \equiv 0,$$

i.e.,

$$a(n + 1) = \frac{n + b + 1}{n + 1}a(n) \ \Rightarrow a(n) = \frac{(n + b)!}{n!}C,$$

for some constant independent of n, and plugging in $n = 0$ yields that $1 = a(0) = b!\,C$ and hence $C = 1/b!$. We have just discovered, and proved at the same time, the Chu–Vandermonde identity.

Note that once we have found the eliminated operator $S(N, n)$ and the corresponding $\bar{R}$ in (9.4.1) above, we can present the proof without mentioning how we obtained it. In this case, $\bar{R} = Nk^2$, so in the above notation

$$G(n, k) = -\bar{R}F(n, k) = -Nk^2 F(n, k) = \frac{-(n+1)!\,b!}{(k-1)!^2(n-k+1)!\,(b-k)!}.$$

Now all we have to present are $S(N, n)$ and $G(n, k)$ above and ask you to believe or prove for yourselves the purely routine assertion that

$$S(N, n)F(n, k) = G(n, k+1) - G(n, k).$$

Dixon's identity by elimination

We will now apply the elimination procedure to derive and prove Dixon's celebrated identity of 1903 [Dixo03]. It states that

$$\sum_k (-1)^k \binom{n+a}{n+k}\binom{n+b}{b+k}\binom{a+b}{a+k} = \frac{(n+a+b)!}{n!\,a!\,b!}.$$

Equivalently,

$$\sum_k \frac{(-1)^k}{(n+k)!(n-k)!(b+k)!(b-k)!(a+k)!(a-k)!} = \frac{(n+a+b)!}{n!a!b!(n+a)!(n+b)!(a+b)!}.$$

Calling the summand on the left $F(n, k)$, we have

$$\frac{F(n+1, k)}{F(n, k)} = \frac{1}{(n+k+1)(n-k+1)},$$

$$\frac{F(n, k+1)}{F(n, k)} = \frac{(-1)(n-k)(b-k)(a-k)}{(n+k+1)(b+k+1)(a+k+1)}.$$

It follows that $F(n, k)$ is annihilated by the operators

$$P = N(n+k)(n-k) - 1; \quad Q = K(n+k)(a+k)(b+k) + (n-k)(a-k)(b-k).$$

Rewrite P and Q in descending powers of k, modulo $K - 1$:

$$P = -Nk^2 + (Nn^2 - 1),$$
$$Q = 2(n+a+b)k^2 + 2nab + (K-1)((n+k)(a+k)(b+k)).$$

Now eliminate k^2 to get the following operator that annihilates $F(n, k)$:

$$2(n+a+b+1)P + NQ = 2(n+a+b+1)(Nn^2 - 1) + N(2nab)$$
$$+ (K-1)(N(n+k)(a+k)(b+k)),$$

which equals

$$N[2n(n+a)(n+b)] - 2(n+a+b+1) + (K-1)(N(n+k)(a+k)(b+k)).$$

In the above notation we have found that the k-free operator

$$S(N,n) = N[2n(n+a)(n+b)] - 2(n+a+b+1)$$
$$= 2(n+1)(n+a+1)(n+b+1)N - 2(n+a+b+1)$$

annihilates $a(n) := \sum_k F(n,k)$.

Also,

$$\bar{R}(N,K,n,k) = (N(n+k)(a+k)(b+k))$$

and

$$G(n,k) = -\bar{R}F(n,k) = \frac{(-1)^{k-1}}{(n+k)!(n+1-k)!(b+k-1)!(b-k)!(c+k-1)!(c-k)!}.$$

Once we have found $S(N,n)$ and $G(n,k)$ all we have to do is present them and ask readers to verify that

$$S(N,n)F(n,k) = G(n,k+1) - G(n,k).$$

Nobuki Takayama has developed a software package for handling elimination, using Gröbner bases.

9.5 Discrete holonomic functions

A discrete function $F(m_1,\ldots,m_n)$ is *holonomic* if it satisfies "as many linear recurrence equations (with polynomial coefficients) as possible" without vanishing identically. This notion is made precise in [Zeil90a], and [Cart91]. An amazing theorem of Stafford [Staf78, Bjor79] asserts that every holonomic function can be described in terms of only two such equations that generate it.

In practice, however, we are usually given n equations, one for each of the variables, that are satisfied by F. They can take the form

$$\sum_{j=0}^{L} a_j^{(i)}(m_1,\ldots,m_n)F(m_1,\ldots,m_{i-1},m_i+j,m_{i+1},\ldots,m_n) = 0.$$

In operator notation, this can be rewritten as

$$P^{(i)}(E_{m_i},m_1,\ldots,m_n)F = 0.$$

Now suppose that we want to consider

$$a(m_1, \ldots, m_{n-1}) := \sum_{m_n} F(m_1, \ldots, m_n).$$

By eliminating m_n from the n operators $P^{(i)}$, $i = 1, \ldots, n$, and setting $E_{m_n} = I$ as before, we can obtain $n - 1$ operators $Q^{(i)}(E_{m_i}, m_1, , \ldots, m_{n-1})$, $i = 1, \ldots, n - 1$, that annihilate a. Hence a is holonomic in all its variables. Continuing, we see that summing a holonomic function with respect to any subset of its variables gives a holonomic function in the surviving variables.

9.6   Elimination in the ring of operators

A more general scenario is to evaluate a multiple sum/integral

$$a(\mathbf{n}, \mathbf{x}) := \int_{\mathbf{y}} \sum_{\mathbf{k}} F(\mathbf{n}, \mathbf{k}, \mathbf{x}, \mathbf{y}) dy, \qquad (9.6.1)$$

where F is holonomic in *all* of its variables, both discrete and continuous. Here $\mathbf{n} = (n_1, \ldots, n_a)$, $\mathbf{k} = (k_1, \ldots, k_b)$ are discrete multi-variables, while $\mathbf{x} = (x_1, \ldots, x_a)$, $\mathbf{y} = (y_1, \ldots, y_d)$ are continuous multi-variables.

A function $F(x_1, \ldots, x_r, m_1, \ldots, m_s)$ is holonomic if it satisfies "as many as possible" linear recurrence-differential equations with polynomial coefficients. This is true, in particular, if there exist operators

$$P^{(i)}(D_{x_i}, x_1, \ldots, x_r, m_1, \ldots, m_s) \quad (i = 1, \ldots, r),$$
$$P^{(j)}(E_{m_j}, x_1, \ldots, x_r, m_1, \ldots, m_s) \quad (j = 1, \ldots, s),$$

that annihilate F. By repeated elimination it is seen that if F in (9.6.1) is holonomic in all its variables, so is $a(\mathbf{n}, \mathbf{x})$.

9.7   Beyond the holonomic paradigm

Many combinatorial sequences are not P-recursive (holonomic). The most obvious one is $\{n^{n-1}\}_1^\infty$, which counts rooted labeled trees, and whose exponential generating function

$$T(x) = \sum_{n=1}^{\infty} \frac{n^{n-1}}{n!} x^n$$

satisfies the *transcendental* equation (i.e., the "algebraic equation of infinite degree")

$$T(x) = xe^{T(x)},$$

or, equivalently, the *non-linear* differential equation

$$xT'(x) - T(x) - xT(x)T'(x) = 0.$$

Other examples are $p(n)$, the number of partitions of an integer n, whose ordinary generating function "looks like" a rational function:

$$\sum_{n=0}^{\infty} p(n)x^n = \frac{1}{\prod_{i=1}^{\infty}(1 - x^i)},$$

albeit its denominator is of "infinite degree." The partition function $p(n)$ itself satisfies a recurrence with *constant* coefficients, but, once again, of infinite order (which, however, enables one to compute a table of $p(n)$ rather quickly):

$$\sum_{j=-\infty}^{\infty} (-1)^j p(n - (3j^2 + j)/2) = 0.$$

Let us just remark, however, that the generating function of $p(n)$ is a *limiting formal power series* of the generating function for $p(n,k)$, the number of partitions of n with at most k parts, which is

$$f_k(q) = \sum_{n=0}^{\infty} p(n,k)q^n = \frac{1}{\prod_{i=1}^{k}(1 - q^i)}.$$

These, for each fixed k, are rational functions, and the sequence $f_k(q)$ itself is q-holonomic in k.

Another famous sequence that fails to be holonomic is the sequence of Bell numbers $\{B_n\}$ whose exponential generating function is

$$\sum_{n=0}^{\infty} \frac{B_n}{n!}x^n = e^{e^x - 1},$$

and which satisfies the "infinite order" linear recurrence, with non-polynomial (in fact holonomic) coefficients:

$$B_{n+1} = \sum_{k=0}^{n} \binom{n}{k} B_k.$$

When we say "infinite" order we really mean "indefinite": you need all the terms up to B_n in order to find B_n itself.

There are examples of discrete functions of two variables $f(n,k)$ that satisfy only one linear recurrence equation with polynomial coefficients, like the Stirling numbers of both kinds.

To go beyond the holonomic paradigm, we should be more liberal and allow these more general sequences. But in order to have an algorithmic proof theory, we must, in each case, convince ourselves that the set of equations used to define a sequence (function) well-defines it (with the appropriate initial conditions), and that the class is closed under multiplication and definite summation/integration with respect to (at least) some subsets of the variables.

A general, fully rigorous theory still needs to be developed, and Sheldon Parnes [Parn93] has made important progress towards this goal. Here we will content ourselves with a few simple classes, just to show what we mean.

Consider, for example, the class of functions $F(x, y)$ that satisfy equations of the form

$$P(x, e^x, y, e^y, D_x, D_y)F(x, y) \equiv 0.$$

If $F(x, y, z)$ satisfies three independent equations

$$P_i(x, y, z, e^x, e^y, e^z, D_x, D_y, D_z)F = 0, \qquad (i = 1, 2, 3)$$

then we should be able to eliminate both z and e^z to get an equation

$$R(x, y, e^x, e^y, D_x, D_y, D_z)F(x, y, z) = 0,$$

from which would follow that if

$$a(x, y) := \int F(x, y, z)dz$$

vanishes suitably at $\pm\infty$ then it satisfies a differential equation:

$$R(x, y, e^x, e^y, D_x, D_y, 0)a(x, y) = 0.$$

When we do the elimination, we consider $x, y, z, e^x, e^y, e^z, D_x, D_y, D_z$ as "indeterminates" that generate the algebra

$$K\langle x, y, z, e^x, e^y, e^z, D_x, D_y, D_z \rangle,$$

under the commutation relations

$$D_x x = xD_x + 1, \qquad D_y y = yD_y + 1, \qquad D_z z = zD_z + 1,$$

and

$$D_x e^x = e^x D_x + e^x, \quad D_y e^y = e^y D_y + e^y, \quad D_z e^z = e^z D_z + e^z,$$

where all of the other $\binom{9}{2} - 6$ pairs mutually commute.

9.8　Bi-basic equations

Another interesting example is that of *bi-basic* q-series, which really do occur in "nature" (see [GaR91]). Let us define them precisely. First, recall that a sequence $a(k)$ is q-hypergeometric if $a(k+1)/a(k)$ is a rational function of (q^k, q). A sequence $a(k)$ is bi-basic (p,q)-hypergeometric if $a(k+1)/a(k)$ is a rational function of (p, q, p^k, q^k).

We can no longer expect that a sum like

$$a(n) = \sum_k \binom{n}{k}_p \binom{n}{k}_q$$

will be (p,q)-hypergeometric (unless some miracle happens). If the summand $F(n,k)$ is (p,q)-hypergeometric in both n and k, it means that we can find operators

$$A(p^k, p^n, q^k, q^n, N), \quad B(p^k, p^n, q^k, q^n, K)$$

that annihilate the summand $F(n,k)$. Alas, in order to get an operator $C(p^n, q^n, N)$ annihilating the sum $a(n) = \sum_k F(n,k)$, we need to eliminate both indeterminates p^k and q^k, which is impossible, in general.

The best that we can hope for, in general, is to deal with sums like

$$a(m,n) = \sum_k \binom{m}{k}_p \binom{n}{k}_q$$

and look for one partial (linear) recurrence $R(p^m, q^n, M, N)a(m,n) = 0$. This goal can be achieved (at least generically, i.e., if the summand $F(m,n,k)$ is "(p,q)-holonomic" in the analogous sense).

If $F(m,n,k)$ is (p,q)-holonomic, then it is annihilated by operators

$$A(p^m, q^m, p^n, q^n, p^k, q^k, M), \ B(p^m, q^m, p^n, q^n, p^k, q^k, N),$$

and $C(p^m, q^m, p^n, q^n, p^k, q^k, K)$. It should be possible to eliminate the two variables p^k and q^k to get an operator $R(p^m, q^n, M, N)$ such that for some operators A', B', C', D', we have

$$R = A'A + B'B + C'C + (K-1)D',$$

and hence R annihilates $a(m,n)$. Nobuki Takayama's package [Taka92] should be able to handle such elimination. However it seems that the time and especially memory requirements would be excessive.

9.9   Creative anti-symmetrizing

The ideas in this section are due to Peter Paule, who has applied them very dramatically in the q-context [Paul94].

The method of creative telescoping, described in Chapter 6, uses the obvious fact that

$$\sum_k (G(n, k+1) - G(n, k)) \equiv 0,$$

provided $G(n, \pm\infty) = 0$. So, given a closed form summand $F(n, k)$, it made sense to look for a recurrence operator $P(N, n)$, and an accompanying certificate $G(n, k)$ (which turned out to be always of the form RATIONAL$(n, k)F(n, k)$) such that

$$P(N, n)F(n, k) = G(n, k+1) - G(n, k).$$

This enabled us, by summing over k, to deduce that

$$P(N, n)(\sum_k F(n, k)) = 0.$$

There is another obvious way for a sum to be identically zero. If the summand $F(n, k)$ is *anti-symmetric*, i.e., $F(n, k) = -F(n, n+\alpha - k)$, for some integer α, then $a(n) := \sum_k F(n, k)$ is identically zero. For example,

$$\sum_k \binom{n}{k}^3 (k^3 - (n - k)^3) = 0,$$

for the above trivial reason. If we were to try to apply the program ct to that sum, we would get a certain *first*-order recurrence that would imply the identity once we verify the initial value $n = 0$, but that would be overkill, and, besides, it wouldn't give us the *minimal-order* recurrence.

This suggests the following improvement. Suppose that $F(n, k)$ is anti-symmetric. Instead of looking for $P(N, n)$ and $G(n, k)$ such that

$$H(n, k) := P(N, n)F(n, k) - (G(n, k+1) - G(n, k))$$

is identically zero, it suffices to insist that it be antisymmetric, i.e., that $H(n, k) = -H(n, n - k)$. This idea is yet to be implemented.

This idea is even more promising for multisums. Recall that the fundamental theorem of algorithmic proof theory [WZ92a] asserts that for every hypergeometric term $F(n; k_1, \ldots, k_r)$ there exist an operator $P(N, n)$ and certificates $G_1(n; k), \ldots, G_r(n; k)$ such that

$$H(n; k_1, \ldots, k_r) := P(N, n)F(n; k) - \sum_{i=1}^{r} \Delta_{k_i} G_r(n; k)$$

is identically zero. Suppose that $F(n; k_1, \ldots, k_r)$ is symmetric w.r.t. all permutations of the k_i's. Then it clearly suffices for $H(n; k_1, \ldots, k_r)$ to satisfy the weaker property that its symmetrizer

$$\bar{H}(n; k_1, \ldots, k_r) := \sum_{\pi \in S_r} H(n; k_{\pi(1)}, \ldots, k_{\pi(r)})$$

vanishes identically, since then we would have

$$r! \, P(N, n) \sum_{k_1, \ldots, k_r} F(n; k) = P(N, n) \sum_k \sum_{\pi \in S_r} F(n; \pi(k))$$

$$= \sum_{\pi \in S_r} \sum_k P(N, n) F(n; \pi(k)) = \bar{H}(n; k) = 0.$$

9.10   Wavelets

The Fourier transform decomposes functions (or "signals", in engineer-speak) into linear combinations of exponentials. The exponential function (and its two offspring, the sine and the cosine) satisfies very simple linear differential equations ($f'(x) = f(x)$, and $f''(x) = -f(x)$). It turns out that for some applications it is useful to take *wavelets* as basic building blocks. The Daubechies wavelets [Daub92] are based on *dilation* equations, which are equations of the form

$$\phi(x) = \sum_{k=0}^{L} c_k \phi(2x - k).$$

This motivates introducing a new class of functions, which let's temporarily call "P-Di functions," that are solutions of equations of the form

$$\sum_{k,j=0}^{L} c_{k,j}(x) \phi(2^j x - k) = 0,$$

where the $c_{k,j}$'s are *polynomials* in x. Introduce the "doubling operator" T_x by

$$T_x f(x) := f(2x).$$

Then P-Di functions are exactly those functions $\phi(x)$ that are annihilated by operators of the form $P(T_x, E_x, x)$, where E_x is the shift operator in x: $E_x f(x) := f(x+1)$.

It is easy to see that the class of P-Di functions forms an algebra. Also the ring of operators in T_x, E_x, x forms an associative algebra subject to the "commuting relations"

$$T_x x = 2x T_x, \quad E_x x = x E_x + E_x, \quad E_x T_x = T_x E_x^2.$$

We don't have to stop at one variable. Consider functions $F(x, y)$ that are "Di-holonomic," i.e., satisfy a system of *two independent* (in some sense, yet to be made precise) equations

$$P(x, T_x, E_x, y, T_y, E_y)F(x, y) = 0, \quad Q(x, T_x, E_x, y, T_y, E_y)F(x, y) = 0.$$

Then it should be possible to perform elimination in the ring $K\langle x, T_x, E_x, y, E_y, T_y \rangle$ to eliminate y, getting an operator $R(x, T_x, E_x, T_y, E_y)$, free of y. Now, using the obvious facts that

$$\int_{-\infty}^{\infty} F(x, y+1)dy = \int_{-\infty}^{\infty} F(x, y)dy, \quad \int_{-\infty}^{\infty} F(x, 2y)dy = \frac{1}{2}\int_{-\infty}^{\infty} F(x, y)dy,$$

we immediately see that

$$a(x) := \int_{-\infty}^{\infty} F(x, y)dy$$

is annihilated by the operator $R(x, T_x, E_x, 1/2, 1)$, obtained by substituting 1 for E_x and $1/2$ for T_x. Hence $a(x)$ is a P-Di function of a single variable.

We can further generalize by also allowing differentiations D_x, D_y and considering the corresponding class of operators and functions, allowing also tripling operators, etc. But let's stop here.

9.11   Abel-type identities

Some obvious identities do not fall under the holonomic umbrella. The most obvious one that comes to mind is

$$\sum_{k=0}^{n} \binom{n}{k} n^k = (n+1)^n.$$

Neither the summand $F(n, k) = \binom{n}{k}n^k$ nor the right side is holonomic (why? because $F(n, k)$ is holonomic in k but *not* in n). So the WZ methodology would not seem to work on this sum. However, this is obviously the special case $x = n$ of the more general, and *holonomic* identity,

$$\sum_{k=0}^{n} \binom{n}{k} x^k = (x+1)^n.$$

Thus some non-holonomic identities are specializations of holonomic ones, and one should *chercher la généralisation*, which is not always as easy as in the example above.

Another class of identities that seem to defy the holonomic paradigm, is that of Abel-type identities (see [GKP89], Section 5.4) whose natural habitat appeared to be convolution and Lagrange inversion. Take, for example,

$$\sum_{k=0}^{n} \binom{n}{k} (k+1)^{k-1} (n-k+1)^{n-k} = (n+2)^n. \tag{9.11.1}$$

The summand $F(n,k)$ is neither holonomic in n nor in k, and the right side is not holonomic either. But (9.11.1) is really a specialization of

$$\sum_{k=0}^{n} \binom{n}{k} (k+r)^{k-1} (s-k)^{n-k} = \frac{(r+s)^n}{r}. \tag{9.11.2}$$

Here the summand $F(n,k,r,s)$ is not holonomic, i.e., it does not satisfy a maximally overdetermined system of linear difference equations in n,k,r,s. But, by forming $(KR^{-1}F)/F$ and $(KSF)/F$, we get *two* equations from which we can eliminate k, getting an operator $\Omega(n,r,s,N,R,S)$ annihilating the sum. Then we just check that Ω annihilates the right side and the initial conditions match. See [Maje96].

Let's consider the well-known identity of Euler,

$$\sum_{k} (-1)^k \binom{n}{k} (x-k)^n = n!. \tag{9.11.3}$$

It has many proofs, but let's try to find a proof by elimination. Let $F(n,k,x)$ be the summand on the left, and let $a(n,x)$ be the whole left side. $F(n,k,x)$ is not holonomic in k and x separately, but is in $x-k$. In other words, $F(n,k+1,x+1)/F(n,k,x)$ is a rational function of (n,k,x). F is obviously holonomic in n, and we have

$$\frac{F(n,k+1,x+1)}{F(n,k,x)} = \frac{k-n}{k+1}, \quad \frac{F(n+1,k,x)}{F(n,k,x)} = \frac{(n+1)(x-k)}{n-k+1}.$$

Using the shift operators $NF(n,k,x) := F(n+1,k,x)$, $KF(n,k,x) := F(n,k+1,x)$, and $XF(n,k,x) := F(n,k,x+1)$, the above can be written in operator notation as follows:

$$P_1 F(n,k,x) \equiv 0, \quad P_2 F(n,k,x) \equiv 0,$$

where

$$P_1 = (k+1)KX + (n-k), P_2 = (n-k+1)N - (n+1)(x-k).$$

Let's write P_1 and P_2 in decreasing powers of k (modulo $(K-1)$):

$$P_1 = k(X-1) + n + (K-1)kX,$$
$$P_2 = k(n+1-N) + (n+1)(N-x).$$

Eliminating k, modulo $(K-1)$, we find that the following operator Q also annihilates $F(n,k,x)$:

$$
\begin{aligned}
Q &:= (n+1-N)P_1 - (X-1)P_2 \\
&= (n+1-N)n - (X-1)(n+1)(N-x) + (K-1)(n+1-N)kX \\
&= -(n+1)(XN - n - (x+1)X + x) + (K-1)(n+1-N)kX.
\end{aligned}
$$

It follows that $a(n,x) := \sum_k F(n,k,x)$ is annihilated by the operator

$$XN - n - (x+1)X + x.$$

In everyday parlance, this means that

$$a(n+1, x+1) = na(n,x) + (x+1)a(n, x+1) - xa(n,x).$$

Now we can prove by induction on n that $a(n,x) = n!$ for all x. Obviously $a(0,x) = 1$ for all x. From the above recurrence taken at $x-1$ we have $a(n+1,x) = (n-x+1)a(n,x-1) + xa(n,x)$, which, by inductive hypothesis, is $(n-x+1+x)n! = (n+1)!$, and we are done. Identities of Abel type have been studied by John Majewicz in [Maje96], and by Ekhad and Majewicz in [EkM96], where they give a short, WZ-style proof of Cayley's formula for counting labeled trees.

9.12 Another semi-holonomic identity

Consider problem 10393 in the *American Mathematical Monthly* (**101** (1994), p. 575 (June-July issue)), proposed by Jean Anglesio. It asks us to prove that

$$\int_0^\infty \frac{e^{-ax}(1-e^{-x})^n}{x^r}\,dx = \frac{(-1)^r}{(r-1)!} \sum_{k=0}^n \binom{n}{k}(-1)^k(a+k)^{r-1}\log(a+k).$$

Neither side is completely holonomic in all its variables (why?), but the identity is easily proved by verifying that both sides satisfy the system of partial difference/differential equations, and initial conditions

$$(N-1+A)F = 0, \quad (\frac{\partial}{\partial a} + R^{-1})F = 0,$$

$$((r-1) - rAR^{-1}N^{-1})F(0,r,r) = 0, \qquad F(a,1,1) = \log a - \log(a+1),$$

in which A, R, N are the forward shift operators in a, r, n, respectively.

9.13   The art

Until now, we have discussed only the *science* of identities. We conclude here with a very brief mention of some of the great *art* that has been created in this area. Many identities in combinatorics are still out of the range of computers, but even if one day they would all be computerizable, that would by no means render them obsolete, since the *ideas* behind the human proofs are often much more important than the theorems that are being proved.

Often identities are tips of icebergs that lead to beautiful depths. For example, the Macdonald identities [Macd72] led Victor Kac [Kac 85, p. xiii] to the discovery of the representation theory of Kac-Moody algebras. Another example is Rota's Umbral Calculus [RoR78, Rom84] which started out as an attempt to unify and explain Sheffer-type identities and to rigorize the 19th-century umbral methods. This theory turned out to have a life of its own, and its significance and depth far transcends the identities that it tried to explain.

Yet another example is the theory of *species*, that was launched by Joyal [Joya81]. This too was initially motivated by identities between formal power series and the formula of Cauchy for the number of labeled trees. It is now a flourishing theory at the hands of Francois Bergeron, Gilbert Labelle, Pierre Leroux [BeLL94] and many others. Its depth and richness far surpasses the sum of its truths, most of which are identities.

To paraphrase a famous saying of Richard Askey [Aske84]:

> There are many identities and no single way of looking at them can illuminate all of them or even all the important aspects of one example.

A good case in point is the healthy rivalry between representation theory and combinatorics. Take for example, the celebrated Cauchy identity:

$$\prod_{1 \le i,j \le n} \frac{1}{1 - x_i y_j} = \sum_\lambda s_\lambda(x_1, \dots, x_n) s_\lambda(y_1, \dots, y_n).$$

The proof of it is a mere exercise in high school algebra (e.g., [Macd95], I.4, ex. 6). However, both the representation theory approach to its proof (that led to Roger Howe's extremely fruitful concept of "dual pairs"), and Knuth's [Knut70] classical bijective proof, that led to a whole branch of bijective combinatorics, contributed so much to our mathematical culture.

Another breathtaking combinatorial theory, that led to the discovery and insightful proofs of many identities, is the so-called *Schützenberger methodology*, sometimes called the *Dyck-Schützenberger-Viennot* (DSV) approach. It was motivated by formal languages and context-free grammars, and proved particularly useful in combinatorial problems that arose in statistical physics [Vien85]. It is vigorously pursued by the École Bordelaise (e.g., [Bous91]).

The last two examples are also connected with the name of Dominique Foata. The combinatorial proof of identities like

$$\binom{a+b}{a+k}\binom{a+c}{c+k}\binom{b+c}{b+k} = \sum_n \frac{(a+b+c-n)!}{(a-n)!\,(b-n)!\,(c-n)!\,(n+k)!\,(n-k)!},$$

was one of the inspirations for the very elegant and extremely influential Cartier-Foata [CaFo69] theory of the commutation monoid. While the above identity (essentially the Pfaff-Saalschütz identity) and all the other binomial-coefficient identities proved there can now be done by our distinguished colleague Shalosh B. Ekhad, as the reader can check with the package EKHAD described in Appendix A below, no computer would ever (or at least for a very long time to come) develop such a beautiful *theory* and such beautiful human *proofs* that are much more important than the theorems they prove. Furthermore, the Cartier-Foata theory, in its geometric incarnation via Viennot's theory of heaps [Vien86], had many successes in combinatorial physics and animal-counting.

Finally, we must mention the combinatorial revolution that took place in the theory of special functions. It was Joe Gillis who made the first connection [EvGi76]. Combinatorial special function theory became a full-fledged research area with Foata's [Foat78] astounding proof of the *Mehler formula* [Rain60, p. 198, Eq. (2)]

$$\sum_{n=0}^{\infty} \frac{H_n(x)H_n(y)t^n}{n!} = (1-4t^2)^{-\frac{1}{2}} \exp\left[y^2 - \frac{(y-2xt)^2}{1-4t^2} \right],$$

where $H_n(x)$ are the Hermite polynomials. While this formula too is completely automatable nowadays (see [AlZe90], or do

```
AZpapc(n!*(1-4*t**2)**(-1/2)*exp(y**2-(y-2*x*t)**2/(1-4*t**2))/ t**(n+1),t,x);
```

in EKHAD), it is lucky that it was not so back in 1977, since it is possible that knowing that the Mehler identity is routine would have prevented Dominique Foata from trying to find another proof. What emerged was [Foat78], the starting point for a very elegant and fruitful combinatorial theory of special functions [Foat83, Stre86, Zeng92]. The *proofs* and the *theory* here (as elsewhere) are far more important than the identities themselves.

On the other hand, formulas like Macdonald's, Mehler's or Saalschütz's could have been discovered and first proved, by computer. Let's hope that in the future, computers will supply us humans with many more beautiful identities, that will turn out to be tips of many beautiful icebergs to come. So the moral is that we need both tips and icebergs, since tips by themselves are rather boring (but not the activity of looking for them!), and icebergs are nice, but we would never find them without their tips.

9.14   Exercises

1. Using the elimination method of Section 9.4, find a recurrence satisfied by

$$a(n) := \sum_k \binom{n-k}{k}.$$

(No credit for other methods!)

2. Find a recurrence satisfied by

$$a(n) := \sum_k \binom{n}{k}\binom{n+k}{k}.$$

3. Using the method of Section 9.4, evaluate, if possible, the following sum:

$$a(n) := \sum_k \frac{(a+k-1)!\,(b+k-1)!\,(c-a-b+n-k-1)!}{k!\,(n-k)!\,(c+k-1)!}.$$

If you succeed you will have rediscovered and reproved the Pfaff–Saalschütz identity.

Appendix A

The WWW sites and the software

Several programs that implement the algorithms in this book can be found on the diskette that comes with the book, as well as on the WorldWideWeb. The programs are of two kinds: some Maple programs and some Mathematica programs. It should be noted at once that both the individual programs and the packages in their entirety will continue to evolve after the publication of this book. Readers are advised to consult from time to time the WorldWideWeb pages that we have created for this book, so as to update their packages as updates become available. These pages will be maintained at two sites (URL's):

> http://www.cis.upenn.edu/~wilf/AeqB.html

and

> http://www.math.rutgers.edu/~zeilberg/programsAB.html

The Maple programs are in packages EKHAD and qEKHAD. The Mathematica programs are Gosper, Hyper, and WZ. We describe these individually below. On our WWW page there are links to other programs that are cited in this book, such as the Mathematica implementation Zb.m of the creative telescoping algorithm, by Peter Paule and Markus Schorn, and the Hyp package of C. Krattenthaler. The Paule-Schorn programs can be obtained from

> http://www.risc.uni-linz.ac.at/research/combinat/risc
> /software/PauleSchorn/

(or else from the link on the home page of this book). Krattenthaler's programs are available from

> http://www.mat.univie.ac.at/~kratt/hyp_hypq/hyp.html

A.1   The Maple packages EKHAD and qEKHAD

EKHAD is a package of Maple programs. Of these, the ones that are specifically
mentioned in this book are ct, zeil, findrec, AZd, AZc, AZpapc, AZpapd, and
celine. If you enter Maple and give the command read 'EKHAD'; then the package
will be read in. If you then type ezra();, you will see a list of the routines contained
in the package. If you then type ezra(ProcedureName);, you will obtain further
information about that particular procedure. The procedures that are contained in
EKHAD are as follows.

- The program ct implements the method of creative telescoping that is described in Chapter 6 of this book. A call to ct(SUMMAND,ORDER,k,n,N) finds
  a recurrence for SUMMAND, which is a function of the running variable n and
  the summation variable k, in the parameters k and n, of order ORDER. The
  input should be a product of factorials and/or binomial coefficients and/or
  rising factorials, where $(a)_k$ is denoted by rf(a,k), and/or powers of k and
  n, and, optionally, a polynomial factor.

  The output consists of an operator ope(N,n) and a certificate R(n,k) with
  the properties that if we define G(n,k):=R(n,k)*SUMMAND then

  ope(N,n)SUMMAND(n,k)=G(n,k+1)-G(n,k),

  which is a routinely verifiable identity.

  For example, if we make a call to ct(binomial(n,k),1,k,n,N); we obtain
  the output N-2, k/(k-n-1), in which N is always the forward shift operator
  in n. For more information about this program, see Section 6.5 of this book.

- Program zeil can be called in several ways. zeil(SUMMAND,k,n,N,MAXORDER),
  for instance, will produce output as in ct above, except that if the program
  fails to find a recurrence of order 1, it will look for one of order 2, etc., up to
  MAXORDER, which has a default of 6. For the other ways to call this program
  see the internal program documentation.

- Program zeilpap is a verbose version of zeil.

- AZd and AZc, and their verbose versions AZpapd and AZpapc implement the
  algorithms in [AlZe90] that were mentioned above on page 114.

- Program celine may be called by celine(SUMMAND,ii,jj). Its operation
  has been described on page 59 of this book.

The package qEKHAD is similar to EKHAD except that it deals with q-identities. In addition TRIPLE_INTEGRAL.maple, with its associated sample input file inTRIPLE, and DOUBLE_SUM_SINGLE_INTEGRAL.maple are Maple implementations of two important cases of the algorithm of [WZ92a].

A.2   Mathematica programs

- The Gosper program does indefinite hypergeometric summation. After getting into Mathematica, read it in with <<gosper.m. Then

GosperSum[f[k],{k,k0,k1}]

will output the sum

$$\sum_{k=k0}^{k1} f[k]$$

as a hypergeometric term, if there exists such a term, or will return the input sum unevaluated, if no such term exists. Examples of the use of this program begin on page 89 of this book.

- The Hyper program solves recurrence relations with polynomial coefficients, where "solves" means that it will return a solution as a sum of a fixed number of hypergeometric terms, if such a solution exists, or "{}", if no such solution exists. First get into Mathematica, read in Hyper, and type "? Hyper". You will see the following documentation:

  - Hyper[eqn, y[n]] finds at least one hypergeometric solution of the homogeneous equation eqn over the field of rational numbers Q (provided any such solution exists).

  - Hyper[eqn, y[n], Solutions -> All] finds a generating set (not necessarily linearly independent) for the space of solutions generated by hypergeometric terms over Q.

  - Hyper[eqn, y[n], Quadratics->True] finds solutions over quadratic extensions of Q.

  - Solutions y[n] are described by giving their rational consecutive term ratio representations y[n+1]/y[n]. Warning: The worst-case time complexity of Hyper is exponential in the degrees of the leading and trailing coefficients of eqn.

For example, a call to

```
Hyper[f[n+2]-2(n+2) f[n+1]+(n+1) (n+2) f[n]==0,f[n]]
```

yields the output **n+2**. That means that the hypergeometric term for which $f(n+1)/f(n) = n+2$ is a solution, i.e., $f(n) = (n+1)!$ is a solution. On the other hand, the call

```
Hyper[f[n+2]-2(n+2)f[n+1]+(n+1)(n+2)f[n]==0,f[n],Solutions->All]
```

produces the reply

$$\left\{1+n, \frac{(1+n)^2}{n}, 2+n\right\}.$$

Now we know that all possible hypergeometric term solutions are linear combinations of the three terms $n!$, $(n)!^2/(n-1)!$, and $(n+1)!$. These are not linearly independent, since the sum of the first two is the third. Hence all closed form solutions are of the form $(c_1 + c_2 n)n!$. More examples are worked out in the text in Section 8.5.

- The program WZ finds WZ proofs of identities. It was given in full and its usage was described beginning on page 141 of this book.

Program qHyper is a q-analogue of program Hyper. It finds all q-hypergeometric solutions of q-difference equations with rational coefficients. The program can be obtained either from the home page for this book (see above), or directly from http://www.fmf.uni-lj.si/~petkovsek/software.html

Bibliography

[Abra71] Abramov, S. A., On the summation of rational functions, *USSR Comp. Maths. Math. Phys.* **11** (1971), 324–330.

[Abr89a] Abramov, S. A., Problems in computer algebra that are connected with a search for polynomial solutions of linear differential and difference equations. *Moscow Univ. Comput. Math. Cybernet.* no. 3, 63–68. Transl. from *Vestn. Moskov. univ. Ser. XV. Vychisl. mat. kibernet.* no. 3, 56–60.

[Abr89b] Abramov, S. A., Rational solutions of linear differential and difference equations with polynomial coefficients. *U.S.S.R. Comput. Maths. Math. Phys.* **29**, 7–12. Transl. from *Zh. vychisl. mat. mat. fiz.* **29**, 1611–1620.

[Abr95] Abramov, S. A., Rational solutions of linear difference and q-difference equations with polynomial coefficients, in: T. Levelt, ed., *Proc. ISSAC '95*, ACM Press, New York, 1995, 285–289.

[ABP95] Abramov, S. A., Bronstein, M. and Petkovšek, M., On polynomial solutions of linear operator equations, in: T. Levelt, ed., *Proc. ISSAC '95*, ACM Press, New York, 1995, 290–296.

[AbP94] Abramov, S. A., and Petkovšek, M., D'Alembertian solutions of linear differential and difference equations, in: J. von zur Gathen, ed., *Proc. ISSAC '94*, ACM Press, New York, 1994, 169–174.

[APP95] Abramov, S. A., Paule, P., and Petkovšek, M., q-Hypergeometric solutions of q-difference equations, *Discrete Math.*, submitted.

[AlZe90] Almkvist, G., and Zeilberger, D., The method of differentiating under the integral sign, *J. Symbolic Computation* **10** (1990), 571–591.

[AEZ93] Andrews, George E., Ekhad, Shalosh B., and Zeilberger, Doron, A short proof of Jacobi's formula for the number of representations of an integer as a sum of four squares, *Amer. Math. Monthly* **100** (1993), 274–276.

[Andr98] Andrews, George E., Pfaff's method. I. The Mills-Robbins-Rumsey determinant, Selected papers in honor of Adriano Garsia (Taormina, 1994), *Discrete Math.* **193** (1998), 43–60.

[Aske84]   Askey, R. A., *Preface*, in: R.A. Askey, T.H. Koornwinder, and W. Schempp, Eds., *"Special functions: group theoretical aspects and applications"*, Reidel, Dordrecht, 1984.

[BaySti]   Bayer, David, and Stillman, Mike, *Macaulay*, a computer algebra system for algebraic geometry. [Available by anon. ftp to math.harvard.edu].

[Bell61]   Bellman, Richard, *A Brief Introduction to Theta Functions*, Holt, Rinehart and Winston, New York, 1961.

[BeO78]   Bender, E., and Orszag, S.A., *Advanced Mathematical Methods for Scientists and Engineers*, New York: McGraw-Hill, 1978.

[BeLL94]   Bergeron, F., Labelle, G., and Leroux, P., *Théorie des espèces et combinatoire des structures arborescentes*, Publ. LACIM, UQAM, Montréal, 1994.

[Bjor79]   Björk, J. E., *Rings of Differential Operators*, North-Holland, Amsterdam, 1979.

[Bous91]   Bousquet-Mélou, M., *q-énumération de polyominos convexes*, Publications de LACIM, UQAM, Montréal, 1991.

[Bres93]   Bressoud, David, Review of "The problems of mathematics, second edition," by Ian Stewart, *Math. Intell.* **15**, #4 (Fall 1993), 71–73.

[BrPe94]   Bronstein, M., and Petkovšek, M., On Ore rings, linear operators and factorisation, *Programming and Comput. Software* **20** (1994), 14–26.

[Buch76]   Buchberger, Bruno, Theoretical basis for the reduction of polynomials to canonical form, SIGSAM Bulletin **39** (Aug. 1976), 19–24.

[CaFo69]   Cartier, P., and Foata, D., *Problèmes combinatoires de commutation et réarrangements*, Lecture Notes Math. **85**, Springer, Berlin, 1969.

[Cart91]   Cartier, P., Démonstration "automatique" d'identités et fonctions hypergéometriques [d'après D. Zeilberger], Séminaire Bourbaki, exposé n° 746, *Astérisque* **206** (1992), 41–91, SMF.

[Cavi70]   Caviness, B. F., On canonical forms and simplification, *J. Assoc. Comp. Mach.* **17** (1970), 385–396.

[Chou88]   Chou, Shang-Ching, An introduction to Wu's methods for mechanical theorem proving in geometry, *J. Automated Reasoning* **4** (1988), 237–267.

[Cipr89]   Cipra, Barry, How the Grinch stole mathematics, *Science* **245** (August 11, 1989), 595.

[Cohn65]   Cohn, R. M., *Difference Algebra*, Interscience Publishers, New York, 1965.

[Com74]   Comtet, L., *Advanced Combinatorics: The art of finite and infinite expansions*, D. Reidel Publ. Co., Dordrecht-Holland, Transl. of *Analyse Combinatoire, Tomes I et II*, Presses Universitaires de France, Paris, 1974.

[Daub92]   Daubechies, Ingrid, *Ten Lectures on Wavelets*, SIAM, Philadelphia, 1992.

[Davi95] Davis, Philip J., The rise, fall, and possible transfiguration of triangle geometry: a mini-history, *Amer. Math. Monthly* **102** (1995), 204–214.

[Dixo03] Dixon, A. C., Summation of a certain series, *Proc. London Math. Soc.* (1) **35** (1903), 285–289.

[Ekha89] Ekhad, Shalosh B., Short proofs of two hypergeometric summation formulas of Karlsson, *Proc. Amer. Math. Soc.* **107** (1989), 1143–1144.

[Ekh90a] Ekhad, S. B., A very short proof of Dixon's theorem, *J. Comb. Theory, Ser. A*, **54** (1990), 141–142.

[Ekh90b] Ekhad, Shalosh B., A 21st century proof of Dougall's hypergeometric sum identity, *J. Math. Anal. Appl.*, **147** (1990), 610–611.

[Ekh91a] Ekhad, S. B., A one-line proof of the Habsieger–Zeilberger G_2 constant term identity, *Journal of Computational and Applied Mathematics* **34** (1991), 133–134.

[Ekh91b] Ekhad, Shalosh B., A short proof of a "strange" combinatorial identity conjectured by Gosper, *Discrete Math.* **90** (1991), 319–320.

[Ekha93] Ekhad, Shalosh B., A short, elementary, and easy, WZ proof of the Askey–Gasper inequality that was used by de Branges in his proof of the Bieberbach conjecture, Conference on Formal Power Series and Algebraic Combinatorics (Bordeaux, 1991), *Theoret. Comput. Sci.* **117** (1993), 199–202.

[EkM96] Ekhad, S. B., and Majewicz, J. E., A short WZ-style proof of Abel's identity, The Foata Festschrift, *Electron. J. Combin.* **3** (1996) no. 2, #R16.

[EkPa92] Ekhad, S. B., and Parnes, S., A WZ-style proof of Jacobi polynomials' generating function, *Discrete Math.* **110** (1992), 263–264.

[EkTr90] Ekhad, S. B., and Tre, S., A purely verification proof of the first Rogers–Ramanujan identity, *J. Comb. Theory, Ser. A*, **54** (1990), 309–311.

[EvGi76] Even, S. and Gillis, J., Derangements and Laguerre polynomials, *Proc. Cambr. Phil. Soc.* **79** (1976), 135–143.

[Fase45] Fasenmyer, Sister Mary Celine, *Some generalized hypergeometric polynomials*, Ph.D. dissertation, University of Michigan, November, 1945.

[Fase47] Fasenmyer, Sister Mary Celine, Some generalized hypergeometric polynomials, *Bull. Amer. Math. Soc.* **53** (1947), 806–812.

[Fase49] Fasenmyer, Sister Mary Celine, A note on pure recurrence relations, *Amer. Math. Monthly* **56** (1949), 14–17.

[Foat78] Foata, D., A combinatorial proof of the Mehler formula, *J. Comb. Theory Ser. A* **24** (1978), 250–259.

[Foat83] Foata, D., Combinatoire de identités sur les polynômes orthogonaux, *Proc. Inter. Congress of Math.* (Warsaw. Aug. 16-24, 1983), Warsaw, 1984.

[GaR91]   Gasper, George, and Rahman, Mizan, *Basic hypergeometric series*, Encycl. Math. Appl. **35**, Cambridge University Press, Cambridge, 1991.

[Gess95] Gessel, Ira, Finding identities with the WZ method, Symbolic computation in combinatorics Δ_1 (Ithaca, NY, 1993), *J. Symbolic Comput.* **20** (1995), 537–566.

[GSY95]   Gessel, I. M., Sagan, B. E., and Yeh, Y.-N., Enumeration of trees by inversions, *J. Graph Theory* **19** (1995), 435–459.

[Gosp77] Gosper, R. W., Jr., Indefinite hypergeometric sums in MACSYMA, in: *Proc. 1977 MACSYMA Users' Conference*, Berkeley, 1977, 237–251.

[Gosp78] Gosper, R. W., Jr., Decision procedure for indefinite hypergeometric summation, *Proc. Natl. Acad. Sci. USA* **75** (1978), 40–42.

[GKP89]   Graham, Ronald L., Knuth, Donald E., and Patashnik, Oren, *Concrete mathematics*, Addison Wesley, Reading, MA, 1989.

[Horn92] Hornegger, J., *Hypergeometrische Summation und polynomiale Rekursion*, Diplomarbeit, Erlangen, 1992.

[Joya81] Joyal, A., Une théorie combinatoire de series formelles, *Advances in Math.* **42** (1981), 1-82.

[Kac 85] Kac, V., *Infinite dimensional Lie algebras*, 2nd Ed., Cambr. Univ. Press, Cambridge, 1985.

[Knop93] Knopp, Marvin I., *Modular functions*, Second edition, Chelsea, 1993.

[Knut70] Knuth, D. E., Permutation matrices and generalized Young Tableaux, *Pacific J. Math.* **34** (1970), 709-727.

[Koep95] Koepf, W., REDUCE package for indefinite and definite summation, *SIGSAM Bulletin* **29** (1995), 14–30.

[Koor93] Koornwinder, Tom H., On Zeilberger's algorithm and its q-analogue, *J. Comp. Appl. Math* **48** (1993), 91–111.

[LPS93]   Lisonek, P., Paule, P., and Strehl, V., Improvement of the degree setting in Gosper's algorithm, *J. Symbolic Computation* **16** (1993), 243–258.

[Loos83] Loos, R., Computing rational zeroes of integral polynomials by p-adic expansion, *SIAM J. Comp.* **12** (1983), 286–293.

[Macd72] Macdonald, I. G., Affine root systems and Dedekind's η-function, *Invent. Math.* **15** (1972), 91-143.

[Macd95] Macdonald, Ian G., *Symmetric functions and Hall polynomials*, Second edition, Oxford University Press, Oxford, 1995.

[Maje96] Majewicz, J. E., WZ-style certification procedures and Sister Celine's technique for Abel-type sums, *J. Differ. Equations Appl.* **2** (1996), 55–65.

[Man93]   Man, Y. K., On computing closed forms for indefinite summations, *J. Symbolic Computation* **16** (1993), 355–376.

[Mati70] Matijaszevic, Ju. V., Solution of the tenth problem of Hilbert, *Mat. Lapok*, **21** (1970), 83–87.

[MRR87] Mills, W. H., Robbins, D. P., and Rumsey, H., Enumeration of a symmetry class of plane partitions, *Discrete Math.* **67** (1987), 43–55.

[Parn93] Parnes, S., *A differential view of hypergeometric functions: algorithms and implementation*, Ph.D. dissertation, Temple University, 1993.

[PaSc94] Paule, P., and Schorn, M., A Mathematica version of Zeilberger's algorithm for proving binomial coefficient identities, *J. Symbolic Computation*, submitted.

[Paul94] Paule, Peter, Short and easy computer proofs of the Rogers–Ramanujan identities and of identities of similar type, *Electronic J. Combinatorics* **1** (1994), #R10.

[Petk91] Petkovšek, M., *Finding closed-form solutions of difference equations by symbolic methods*, Ph.D. thesis, Carnegie-Mellon University, CMU-CS-91-103, 1991.

[Petk92] Petkovšek, M., Hypergeometric solutions of linear recurrences with polynomial coefficients, *J. Symbolic Computation*, **14** (1992), 243–264.

[Petk94] Petkovšek, M., A generalization of Gosper's algorithm, *Discrete Math.* **134** (1994), 125–131.

[PeW96] Petkovšek, Marko, and Wilf, Herbert S., A high-tech proof of the Mills–Robbins–Rumsey determinant formula, The Foata Festschrift, *Electron. J. Combin.* **3** (1996) no. 2, #R19.

[PiSt95] Pirastu, Roberto, and Strehl, Volker, Rational summation and Gosper-Petkovšek representation, Symbolic computation in combinatorics Δ_1 (Ithaca, NY, 1993), *J. Symbolic Comput.* **20** (1995), 617–635.

[Rain60] Rainville, Earl D., *Special functions*, MacMillan, New York, 1960.

[Rich68] Richardson, Daniel, Some unsolvable problems involving elementary functions of a real variable, *J. Symbolic Logic* **33** (1968), 514–520.

[Risc70] Risch, Robert H., The solution of the problem of integration in finite terms, *Bull. Amer. Math. Soc.* **76** (1970), 605–608.

[Rom84] Roman, S., *The umbral calculus*, Academic Press, New-York, 1984.

[RoR78] Roman, S., and Rota, G. C., The umbral calculus, *Advances in Mathematics* **27** (1978), 95-188.

[Sarn93] Sarnak, Peter, *Some applications of modular forms*, Cambridge Math. Tracts #99, Cambridge University Press, Cambridge, 1993.

[Staf78] Stafford, J. T., Module structure of Weyl algebras, *J. London Math. Soc.*, **18** (1978), 429–442.

[Stan80] Stanley, R. P., Differentiably finite power series. *European J. Combin.* **1** (1980), 175–188.

[Stre86]   Strehl, Volker, Combinatorics of Jacobi configurations I: complete oriented matchings, in: *Combinatoire énumérative*, G. Labelle et P. Leroux eds., LNM **1234**, 294-307, Springer, 1986.

[Stre93]   Strehl, Volker, Binomial sums and identities, Maple Technical Newsletter **10** (1993), 37–49.

[Taka92]   Takayama, Nobuki, An approach to the zero recognition problem by Buchberger's algorithm, *J. Symbolic Computation* **14** (1992), 265–282.

[vdPo79]   van der Poorten, A., A proof that Euler missed... Apéry's proof of the irrationality of $\zeta(3)$, *Math. Intelligencer* **1** (1979), 195–203.

[Verb74]   Verbaeten, P., The automatic construction of pure recurrence relations, *Proc. EUROSAM '74, ACM-SIGSAM Bulletin* **8** (1974), 96–98.

[Vien85]   Viennot, X. G., Problèmes combinatoire posés par la physique statistique, *Séminaire Bourbaki n° 626*, Asterisque **121-122** (1985), 225-246.

[Vien86]   Viennot, X. G., Heaps of pieces I: Basic definitions and combinatorial lemmas, in: *Combinatoire énumérative*, G. Labelle et P. Leroux eds., LNM **1234**, 321-350, Springer, 1986.

[Wilf91]   Wilf, Herbert S., Sums of closed form functions satisfy recurrence relations, unpublished, March, 1991 (available at the WorldWideWeb site `http://www.math.upenn.edu/~wilf/website/other.html`).

[Wilf94]   Wilf, Herbert S., *generatingfunctionology* (2nd ed.), Academic Press, San Diego, 1994.

[WZ90a]   Wilf, Herbert S., and Zeilberger, Doron, Rational functions certify combinatorial identities, *J. Amer. Math. Soc.* **3** (1990), 147–158.

[WZ90b]   Wilf, Herbert S., and Zeilberger, Doron, Towards computerized proofs of identities, *Bull. (N.S.) Amer. Math. Soc.* **23** (1990), 77–83.

[WZ92a]   Wilf, Herbert S., and Zeilberger, Doron, An algorithmic proof theory for hypergeometric (ordinary and "q") multisum/integral identities, *Inventiones Mathematicæ* **108** (1992), 575–633.

[WZ92b]   Wilf, Herbert S., and Zeilberger, Doron, Rational function certification of hypergeometric multi-integral/sum/"q" identities, *Bull. (N.S.) of the Amer. Math. Soc.* **27** (1992), 148–153.

[WpZ85]   Wimp, J., and Zeilberger, D., Resurrecting the asymptotics of linear recurrences. *J. Math. Anal. Appl.* **111** (1985), 162–176.

[WpZ89]   Wimp, J., and Zeilberger, D., How likely is Pólya's drunkard to stay in $x \geq y \geq z$?, *J. Stat. Phys.* **57** (1989), 1129–1135.

[Yen 93]   Yen, Lily, *Contributions to the proof theory of hypergeometric identities*, Ph.D. dissertation, University of Pennsylvania, 1993.

[Yen96]   Yen, Lily, A two-line algorithm for proving terminating hypergeometric identities, *J. Math. Anal. Appl.* **198** (1996), 856–878.

[Yen97]   Yen, Lily, A two-line algorithm for proving q-hypergeometric identities, *J. Math. Anal. Appl.* **213** (1997), 1–14.

[Zeil82]   Zeilberger, Doron, Sister Celine's technique and its generalizations, *J. Math. Anal. Appl.* **85** (1982), 114–145.

[Zeil90a]   Zeilberger, Doron, A holonomic systems approach to special functions identities, *J. of Computational and Applied Math.* **32** (1990), 321–368.

[Zeil90b]   Zeilberger, Doron, A fast algorithm for proving terminating hypergeometric identities, *Discrete Math.* **80** (1990), 207–211.

[Zeil91]   Zeilberger, Doron, The method of creative telescoping, *J. Symbolic Computation* **11** (1991), 195–204.

[Zeil93]   Zeilberger, Doron, Closed form (pun intended!), *A Tribute to Emil Grosswald: Number Theory and Related Analysis*, M. Knopp and M. Sheingorn, eds., Contemporary Mathematics **143**, 579–607, AMS, Providence, 1993.

[Zeil96a]   Zeilberger, Doron, Proof of the alternating sign matrix conjecture, The Foata Festschrift, *Electron. J. Combin.* **3** (1996) no. 2, #R13.

[Zeil96b]   Zeilberger, Doron, If A_n Has $6n$ Dyes in a Box, With Which He Has To Fling [at least] n Sixes, Then A_n Has An Easier Task Than A_{n+1}, at Eaven Luck, *Amer. Math. Monthly* **103** (1996), 265.

[Zeng92]   Zeng, J., Weighted derangements and the linearization coefficients of orthogonal Sheffer polynomials, *Proc. London Math. Soc.* (3) **65** (1992), 1-22.

Index

Printed and bound by CPI Group (UK) Ltd, Croydon, CR0 4YY

23/10/2024

01778245-0001